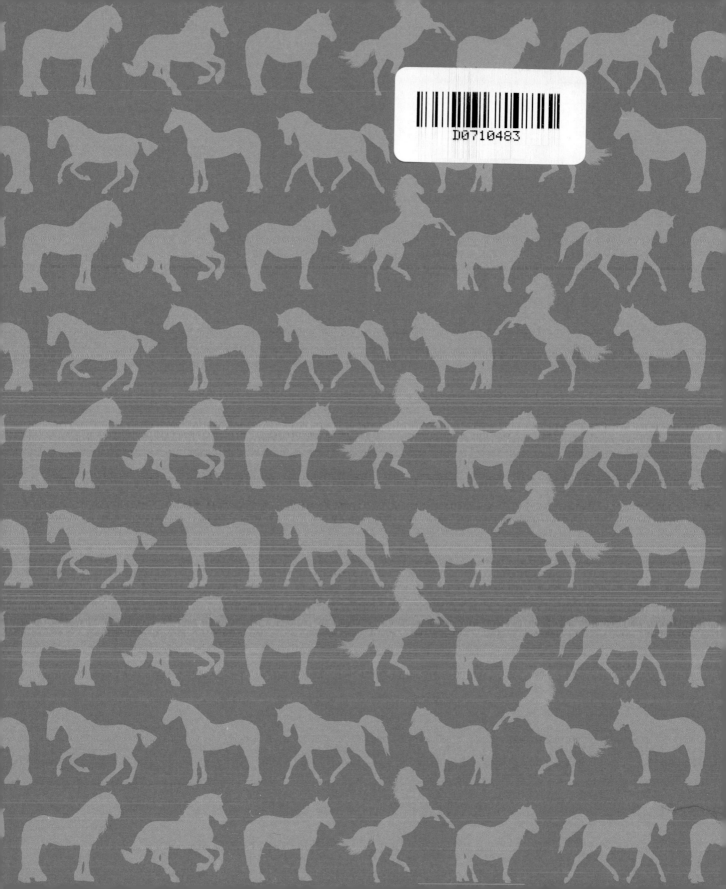

The HORSE

First published in the United States and Canada in 2019 by
Princeton University Press
41 William Street
Princeton, NJ 08540

First published in the United Kingdom in 2019 by
Ivy Press
An imprint of The Quarto Group
The Old Brewery, 6 Blundell Street
London N7 9BH, United Kingdom

Library of Congress Control Number: 2018963947

ISBN: 978-0-691-17877-6

This book was conceived, designed, and produced by
Ivy Press
58 West Street, Brighton BN1 2RA, United Kingdom

Publisher Susan Kelly
Editorial Director Tom Kitch
Art Director James Lawrence
Project Editor Natalia Price-Cabrera
Design JC Lanaway
Picture Researcher Katie Greenwood
Illustrator John Woodcock

Printed in China

10 9 8 7 6 5 4 3 2 1

The HORSE
A Natural History

DEBBIE BUSBY AND CATRIN RUTLAND

PRINCETON UNIVERSITY PRESS

PRINCETON AND OXFORD

Contents ∽

CHAPTER 5
A Directory of Horse Breeds

Appendices

Introducing the Horse ⁓

I t is hard to believe that the ancient ancestors of the horse were four-toed, wild, 2-foot (60-cm)-tall mammals from North America. During the previous 55 million years, the horse has evolved into a truly magnificent animal, one that has lived alongside humans and served many purposes during the last few thousand years. Ancient cave drawings indicate the importance of the horse and little has changed in the present day. Early tapestries, religious texts, historical writings, and stories all show the enormous relevance of horses throughout history and their relationship with man. It has been said that there have been many eras of the horse. These include for consumption (eating), as a utilization and status animal, for herding, for pulling chariots and carriages, in the cavalry, in agriculture, and for leisure. It would be fair to say that horses still undertake most of these tasks, even in the present day, but they have evolved over time. Chariot racing may have gone out of fashion; however, in many parts of the world it is still common to see a bride being taken to her wedding in a horse-drawn carriage or a horse pulling a wagon of goods from village to village. Horses remain a part of the community and still play vital roles in many cultures.

As the horse has grown in size, it has become more graceful and able to achieve greater speeds. However, it has not always been plain sailing for this mighty creature. There were times when it died out in parts of the world. Its closest relatives the tarpan (now extinct) and Przewalski's horse (now endangered) have not fared well, and many breeds of horses are at lower numbers than seen previously. Many of the ancient ancestors and distant relations of the horse have become extinct or endangered. A visit to a natural history museum often reminds us of how many types of horse existed and how many related animals have disappeared with time. Regardless, the number of breeds and types of horses alive today represents a varied and generally healthy population, each with its own characteristics and personalities. From the small Shetland ponies to the great sturdy draft horses, from the fast Thoroughbreds to the gentle therapy ponies, each has its own place in society today.

Right: *Certain characteristics inherent in the Duelmener horse suggest it has a primitive origin. At present, around 300 Duelmener ponies thrive in the Merfelder Burch in Germany, where they lead a semi-wild life searching for food and shelter. Each year, the herd is rounded up and the young stallions are caught and separated from the group. The mares are returned to their habitat with just a couple of stallions.*

Left: *Mounted police in London, UK. The horse continues to have an integral role in human society and an ongoing relationship with humans that shows no signs of abating anytime soon.*

A REWARDING SPECIES

The horse has taken on many roles throughout the years: as a symbol of power and wealth for leaders, kings, and queens in both times of war and peace; as a brave war-horse; and as a gentle pet for young children. The horse has been an athlete from pulling chariots in great Roman amphitheaters to present-day sports, such as racing, show jumping, polo, rodeo, dressage, and hunting. It has been pitted against bulls in fighting arenas alongside human companions, and it has also been a source of food for not only humans but also other animals. Horsehair and leather have been valuable resources over the years and, of course, its gelatin has been used as glue. The horse has shown itself to be strong and steady while

Below: *The English photographer Edweard Muybridge was fascinated by movement, particularly that of the horse. He is well known for his groundbreaking work on animal locomotion in the 1870s.*

pulling a plow or cart, working as a packhorse or down mines, or removing logs from dense forests. For many years, horses have been the most reliable form of transport—for those who could afford it—and, in many parts of the world, this remains true today. Over the course of history, the horse has not only evolved in terms of its anatomy, but also in its relationship with humans and in the types of roles it has played in our lives.

Nowadays, the horse is still used throughout the world as a form of transport, a worker, an athlete, or a much loved pet, and by the military and police forces as an active or show horse.

The horse is also a great muse for the arts, whether we think of the centaur or the unicorn or the many horses that feature

in paintings, photographs, sculptures, literature, movies, and other art forms. Few people can help but feel moved when they think of the famous war-horse stories in books, the movies, or at the theater. Art galleries are filled with paintings and drawings of the majestic horse, often with a proud rider in their finery. Think George Stubbs, Henri de Toulouse-Lautrec, and Franz Marc, to name but a few. There are fairy tales, myths, and legends in abundance from our childhood that depict the adventures of horses, which often involve them coming to the aid of a human or another animals. The horse is a truly mystical creature that has captivated humankind for centuries, and its speed, agility, grace, intelligence, obedience, and beauty have ensured a lasting relationship with humans and made their mark in history.

Above: *Two young girls riding pretend horses. Our childhoods are full of references to the horse, whether in imaginative play, such as this, or in the storybooks we read as children at bedtime. The horse has always been a magical creature and the stuff of myths and legends.*

ABOUT THIS BOOK

This book represents a wealth of knowledge about the horse, collected over the years from archeological findings and historical manuscripts to scientific and veterinary research and practice and social studies. It draws upon research publications as well as detailed anatomical and physiological studies made throughout the years, and it uses information from equine health professionals and owners who know and understand horses.

Chapter 1 explores the evolution and development of the horse. From the first horse, known as the dawn-horse, to the modern-day *Equus ferus caballus*, the history and development of this animal is detailed and discussed. Its ancestors and nearest relatives are investigated and compared, and the journey of the first horse in North America to its worldwide existence is mapped. The transformation of the horse from a wild animal to its present-day domestication and its importance to humans is also explored.

Chapter 2 examines the anatomy and physiology of the horse. We follow the life cycle of the horse from breeding to old age, looking at life as a mammal; reproduction and breeding; diseases and injuries; understanding the skeleton, organs, and vital systems, such as respiration, cardiovascular functions, and the structure of the digestive system. Understanding the equine senses is vital for any owner, therefore, these are explored, and the unique but complex hoof structure and characteristics are detailed alongside modern veterinary and scientific techniques and research, including genetics.

Chapter 3 covers society and behavior. Topics include courtship and mating, birth and parenting, the interactions between different horses and those with humans. Areas such as sleep, communication, play, group and herd dynamics, comprehension, and learning are all covered.

Chapter 4 looks at the relationship between people and horses. It highlights early domestication and the many different uses of horses in historical terms up to the present day. It also details modern-day breeding, economics, and ethics.

Chapter 5 delves into horse breeds, first by introducing the different historic breeds from around the world, and then by looking at individual breeds in detail. This gives an insight into the different types of horses that exist.

Left: *For centuries, humans have cultivated a powerful relationship with the horse. Its roles have been diverse, but one thing that has remained a constant throughout time is the horse's ability to lift our spirits and be a loyal and trusted companion.*

CHAPTER 1

Evolution & Development

Ancestors of the Modern Horse ∽

Much has been learned about the early ancestors of the horse and evolutionary processes from the many fossils that have been discovered. The first ancestor of the modern horse lived 50 to 55 million years ago (mya). The fossil record shows how the horse has changed over time.

EOHIPPUS

The genus *Eohippus* (meaning dawn-horse) has only one species—*Eohippus angustidens*. This horse has also been referred to as *Hyracotherium*, because its skull was harelike in structure.

This now extinct animal has had much written about its size. Early on, it was described as being the size of a fox terrier, and the term "fox size" is still used. The late evolutionary scientist Stephen Jay Gould pointed out that, at about 2 feet (60 cm) tall and 50 pounds (23 kg) in weight, the *Eohippus* was more like a small deer. Although extinct, the fossilized skeletons found in North America show fascinating similarities to and differences from modern equines. With four toes on its front feet, three on the back, and pads on its feet, its limbs were more suitable for walking in soft, ancient forestlands instead of grasslands, and its 44 teeth were more appropriate for eating foliage and fruits.

OROHIPPUS

Around 50 million years ago, there was an evolutionary change into what we now call *Orohippus*. Although its name means "mountain horse," it neither lived in the mountains nor is considered a true horse. Its feet were similar to those of *Eohippus* but the teeth had changed. They had more pronounced crests, dwarfed first premolars, and the last premolar was now a full molar tooth, making grinding actions more possible. *Orohippus* could grind tougher material, so it did not need to rely on eating leaves and fruit. The overall body shape also became slimmer in comparison to *Eohippus*.

Below: *Although the paleontologist Othniel Marsh first described* Eohippus *in 1876, it was later found that* Hyracotherium angustidens, *described by Edward Cope in 1875, was the same species.*

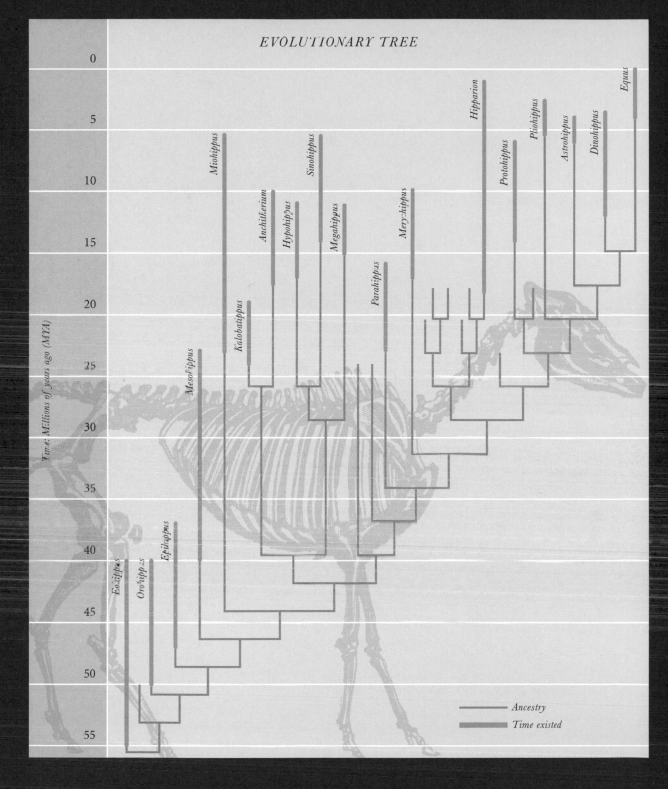

EVOLUTIONARY TREE

Time: Millions of years ago (MYA)

Eohippus
Orohippus
Epihippus
Mesohippus
Miohippus
Kalobatippus
Anchitherium
Hypohippus
Sinohippus
Megahippus
Parahippus
Merychippus
Hipparion
Protohippus
Pliohippus
Astrohippus
Dinohippus
Equus

Ancestry
Time existed

EPIHIPPUS AND MESOHIPPUS

Evolving from *Orohippus*, *Epihippus* continued the evolutionary trend toward eating food that required grinding. Three million years of evolution produced teeth that had developed further and, although it was still only 2 feet (60 cm) tall, that would change a few million years later. About 40 mya, the *Epihippus* had grown in size, and it is now classified as *Mesohippus* (meaning "middle horse"). The North American forests were steadily becoming grasslands. As the environment changed, so did the equids. Although the same height, they were now around 3 feet (90 cm) in length, and these animals actively used three toes on each foot that were more suited for running. Their teeth had further evolved, with a single premolar and six grinding "cheek teeth."

MIOHIPPUS

The *Mesohippus* was not the only equid in existence around this time. The fossils show that a second type of horse coexisted, *Miohippus* ("lesser horse"), around 36 mya. *Miohippus* originally came from *Mesohippus*, with a group splitting off and following a slightly different evolutionary pathway. After four million years of grazing on the planet together, *Miohippus* eventually outlived *Mesohippus*. *Miohippus* was taller and its teeth further evolved. It survived for a substantial amount of time; it was still living 2.5 mya and is considered the direct ancestor of the true equids. Interestingly, it also diversified into two main groups. One remained on the grass plains, but the other became adapted to life in the forests.

HORSE EVOLUTION

From 40 million years ago to 2.6 million years ago, the early horse grew taller and its bone structure changed, for example, from three toes on each foot to just one toe on each foot, as seen on the *Pliohippus*.

MESOHIPPUS
Late Eocene

MERYCHIPPUS
Middle Miocene

PLIOHIPPUS
Late Miocene

OTHER ANCESTORS

After this, the true equines diverge at a greater pace, with many different types of horse seen branching out and starting to spread across the planet. *Kalobatippus* came from the forest-living *Miohippus* and is the probable ancestor of *Anchitherium*, which spread into Asia and then Europe. This in turn evolved into *Sinohippus* in Eurasia and *Hypohippus* and *Megahippus* in North America. The ancestors on the grasslands also evolved into *Parahippus*, which stood at more than 3 feet 3 inches (1 m) tall, had fused leg bones, and now stood on one central toe. It started to resemble the modern horse. Later, the *Merychippus* was abundant and evolved into 16 different grass eaters, the survivors of which are divided into three categories: *Hipparion*, *Protohippus*, and *Pliohippus*. Many of these spread into Europe. It was originally thought that *Pliohippus* was the direct ancestor of the modern horse, but we now know that it gave rise to *Astrohippus*.

It was *Dinohippus* (translated as "terrible horse," 10.3–3.6 mya), endemic across North America, that would develop into *Plesippus* to give the complete history of the modern horse. Named *Plesippus shoshonensis* by paleontologist James W. Gidley in 1930, it was later renamed *Equus simplicidens* after it was realized that earlier specimens were similar. It is often called the Hagerman horse, because a great many fossils have been found in Hagerman, Idaho. This animal certainly resembles the modern horse. It first lived 3.5 mya, was similar in size to a modern Arabian horse, and weighed between 250 and 175 pounds (110–385 kg). It was a sturdy horse, with zebra and donkey-type features.

UNDERSTANDING EVOLUTION

The work invested by early scientists has paved the way for us to understand the equine evolutionary tree. Georges Cuvier (1769–1832) himself, the so-called father of paleontology, studied the first excavated equid fossils in around 1825. In 1839, Richard Owen (1804–92) had described and named the ancestor of all modern horses. In the 1850s and onward, Joseph Leidy (1823–91) drew brilliant equine monographs and had in mind Charles Darwin's (1809–82) great theory of evolution. In the 1870s, Thomas Henry Huxley (1825–95) and Vladimir Kovalevsky (1848–1935) made some revolutionary insights into the ancestry of the horse. Othniel Charles Marsh (1831–99), also a Darwinist, studied the horse in great detail and added much to what was known. In the past two centuries, much has been discovered about these wonderful creatures by many thousands of scientists throughout the world. As anatomical, paleontological, geological, and biological techniques have themselves evolved and become more advanced, so does the insight into the evolution of the modern horse.

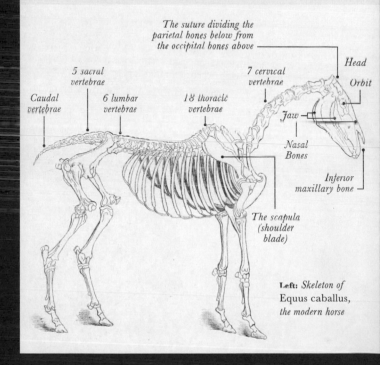

The suture dividing the parietal bones below from the occipital bones above

Head

Orbit

5 sacral vertebrae

7 cervical vertebrae

Caudal vertebrae

6 lumbar vertebrae

18 thoracic vertebrae

Jaw

Nasal Bones

Inferior maxillary bone

The scapula (shoulder blade)

Left: *Skeleton of* Equus caballus, *the modern horse*

Related Species & Distant Cousins ⌾

The modern horse belongs to the species *Equus ferus*, but it is interesting to start further up the family tree to understand how the horse is related to not only the ass and zebra but also the tapir and the rhinoceros.

THE ORDER PERISSODACTYLA

The order Perissodactyla contains a range of species grouped together based on two features: they have either one or three toes on their rear hooves and they are are hindgut fermenters.

There are three extant families in the order: Rhinocerotidae (rhinoceros), Tapiridae (tapirs), and Equidae.

The family Equidae contains all the ancestors of the horse already discussed in the previous section, from *Eohippus* to *Dinohippus*. Today, the only extant genus in Equidae is *Equus*, which contains seven species (see "Family Equidae" below). The following pages will explore the animals in each species, identify subspecies, relate how they evolved and live or lived their lives, and describe some of their unique features.

Above right: *The Somali wild donkey is a subspecies of* Equus africanus, *the forefather of the modern domestic ass. This species is extremely rare both in nature and in captivity.*

Below: *A cladogram showing the evolution of some of the distant cousins of the modern horse.*

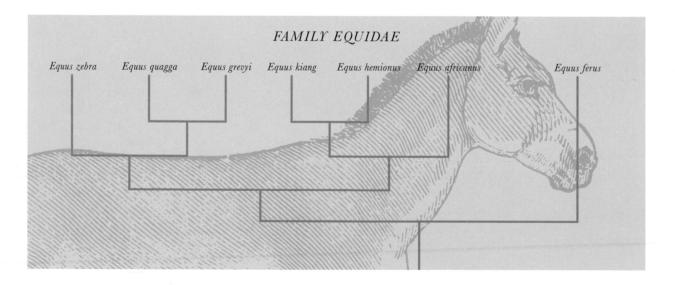

FAMILY EQUIDAE

Equus zebra *Equus quagga* *Equus grevyi* *Equus kiang* *Equus hemionus* *Equus africanus* *Equus ferus*

EQUUS AFRICANUS (AFRICAN WILD ASS)

There are four subspecies of this ass. The first is the Nubian wild ass (*E. a. africanus*), which is probably extinct but may have two small living populations. The last sighting was in the 1970s. It is thought to be the ancestor of the modern donkey, along with another type of wild ass that is presently unknown. DNA evidence

Below: A male (called a jack) and female (a jenny) Equus africanus asinus, the modern donkey or ass. With more than forty million living donkeys on the planet, this domesticated mammal is thriving. If bred with horses (called mules and hinnies) or zebras (zebroid/zonkey/zonkra), the offspring are usually infertile.

suggested this finding in 2010, but later DNA findings also indicated that feral donkeys in Bonaire were most similar to the Nubian ass and different from other asses. The second subspecies is *E. a. somaliensis*, found in Somalia, Eritrea, and Ethiopia. This ass has striped legs, similar to the modern zebra, but is critically endangered, with only around 700 alive in the wild and 200 in captivity. It has a special place within the lineage because, although most domesticated donkeys come from the Nubian wild ass, in Italy the donkeys are descended from this Somalian subspecies. The third subspecies is *E. a. asinus*, the domesticated donkey, which was first domesticated around 3,000 years ago and is now both a companion animal and working animal throughout the world. It is thought that the first domesticated donkeys were reared in Egypt or Mesopotamia. The final subspecies is the now extinct *E. a. atlanticus*. Commonly known as the Atlas or Algerian wild ass, it was much favored by the Romans as a beast to hunt; it is thought to have died out about 300 BCE.

EQUUS FERUS
(WILD HORSE)

The modern horse belongs to the species *Equus ferus*, which contains three subspecies: the domesticated horse (*Equus ferus caballus*), the extinct tarpan (*Equus ferus ferus*), and the endangered Przewalski's horse (*Equus ferus przewalskii*). These are covered on pages 24 to 27.

EQUUS GREVYI
(GRÉVY'S ZEBRA)

This is the largest of the three living zebras, weighing in at up to 100 pounds (450 kg) and standing up to 5 feet 2 inches (1.57 m). It has distinctive narrow stripes and a fairly mulelike look with a large head, rounded ears, and a thick, short neck. Grévy's zebra is found in the wild in Ethiopia and Kenya, although it is endangered. It can breed with the more common plains zebra; fertile offspring are being produced and the two have even been seen herding together. Interestingly, the stallions play an important role in rearing foals. A dominant male will care for several foals together while the mares are eating. Recent reports estimate the current population to be around 2,600 individuals, a steep decline from the already low number of 15,000 in the 1970s. Hunting of the zebra has declined, but a combination of competition for food with other animals, a new plant outgrowing their usual food sources, and loss of habitat still pose a great threat.

Right: *The Indian wild ass (*Equus hemionus khur*) lives in Southern Asia and is also known as the Ghudkhur in the local Gujarati language. It is endangered due to hunting, disease, and habitat decline, but it is gradually on the increase again after conservation efforts.*

Below: *A Grevy's zebra (*Equus Grevyi*), also known as the imperial zebra, nursing her foal in Kenya. This zebra is the largest living wild equid, but it is presently endangered. Thankfully, it can breed successfully with the more common plains zebra.*

EQUUS HEMIONUS (ASIATIC WILD ASS)

The name comes from the Greek and means "half donkey." Although an ass, with notably shorter legs than horses, the onager is fast—it can travel up to 44 mph (70 km/h). It is an ancient equid species, having existed for more than 4 million years. The animals have a black stripe down their back and their coat color changes with the season. There are four extant subspecies: *E. h. hemionus* (Mongolian wild ass), *E. h. kulan* (Turkmenian wild ass), *E. h. onager* (Persian onager or Persian wild ass), and *E. h. khur* (Indian wild ass). A fifth subspecies, *E. h. hemippus* (Syrian wild ass), is now extinct, with the last wild survivor shot in 1927 and the last surviving captive animal dying in the same year.

The four remaining living subspecies are either endangered or near threatened. As with the other endangered species, hunting, habitat loss, and food loss are worrisome, and these animals are also used for traditional medicine and are vulnerable to disease, many of which can be fatal. It should also be noted that the kiang (overleaf) descended from this group and became a different species.

Below: *Turkmenian kulan (*Equus hemionus kulan*) from Central Asia. Despite existing for more than 4 million years, it is endangered; American and Eurasian breeding programs hope to reverse this. It is shorter than a horse, at 3 feet 3 inches to 4 feet 7 inches (100–140 cm) in height.*

EQUUS KIANG
(KIANG)

Its range is now restricted to parts of China (predominantly Tibet), Nepal, India, and Pakistan. It grows to about 4 feet 7 inches (1.4 m) tall, and its woolly fur ensures it is relatively well protected against the elements. It has a stripe running along its back and a coat that changes color with the seasons. There are three subspecies: *E. k. kiang* (western kiang), *E. k. holdereri* (eastern kiang), and *E. k. polyodon* (southern kiang). These animals can live on lowlands, but they can also survive at elevations of 17,000 feet (5,300 m). Because food can be difficult to find in their regions, kiangs eat predominantly grass but also roots, sedges, and shrubs, so they are adaptable. It is perhaps this adaptability, coupled with a lack of predators (only wolves and humans, although hunting kiangs is illegal within the countries they inhabit), that has contributed to their being a "species of least concern" in conservation terms. They inhabit at least 163 locations and have 60,000–70,000 adults within the entire population. The populations are difficult to monitor, because although they do not migrate, they do move with the seasons and food availability.

Above: *The kiang (*Equus kiang*) is the biggest wild ass. In winter, its coat doubles in length for surviving the cold Tibetan climate.*

EQUUS QUAGGA (PLAINS ZEBRA)

Although generally considered to be common, this species was classified as "near threatened" by the IUCN in 2016, and some of the subspecies are critically endangered. In total, there are around 500,000 plains zebras, with about 50 percent being mature adults, but these are not evenly distributed across the subspecies. There are six extant subspecies: *E. q. burchellii* (Burchell's zebra), *E. q. boehmi* (Grant's zebra), *E. q borensis* (maneless zebra), *E. q. chapmani* (Chapman's zebra), *E. q. crawshayi* (Crawshay's zebra), and *E. q. selousi* (Selous' zebra), and the extinct *E. q. quagga*.

The zebra has caused much discussion over the years when trying to classify the species. Because each individual looks so different due to the stripe pattern, it has been difficult to ascertain real differences from perceived differences.

The different subspecies have a number of interesting characteristics; for example, Burchell's zebra has a dorsal line and migrates. Grant's zebra are the smallest at up to 4 feet 6 inches (1.37 m) tall and 660 pounds (300 kg) in weight. The maneless zebra has short hair for a mane and was first described in 1954. Chapman's zebra foals are born with stripes that usually turn black with age, but not always. The stripes often end in brown spots, which give the zebra an unusual look. They are low risk in terms of conservation, and their herd sizes can be immense, many thousands in some cases. Crawshay's zebra has narrow stripes and differing teeth in comparison to other subspecies. Selous' zebra is critically endangered with an estimated 50 individuals remaining.

EQUUS ZEBRA (MOUNTAIN ZEBRA)

Equus zebra contains two subspecies, *E. z. zebra* (Cape mountain zebra) and *E. z. hartmannae* (Hartmann's mountain zebra). It is considered to have a "vulnerable" conservation status, although the positive news is that numbers appear to be increasing due to intense conservation efforts.

Equus ferus ⌒

We have just looked in depth at the seven species within the genus *Equus*. From the asses to the zebra and tarpan, the extinct and those still roaming the planet, these animals give us a greater insight into the domesticated horse.

By far the most common subspecies of *Equus ferus* is the domesticated horse (*E. f. caballus*). There is also the rare wild horse (*E. f. przewalskii*), commonly known as Przewalski's horse, named after the Russio–Polish explorer Nikolay Przhevalsky. Other names for this horse include takhi, dagy, and Dzungarian horse. The third subspecies of *Equus ferus* is the extinct tarpan (*E. f. ferus*).

Below: *The photograph shows one of the last known tarpans (Equus ferus ferus) to exist prior to extinction. It lived at Moscow zoo. It is reported that the last tarpan died in 1909.*

EQUUS FERUS FERUS (TARPAN)

Equus ferus ferus is called the tarpan, which means "wild horse" in Turkish. It is also known as the Eurasian wild horse. The last tarpan died in captivity in 1909 in Russia. This wild horse captivated humans, so in the 1930s several attempts were made to re-create the look of the tarpan by selective breeding. None of these breeds is, of course, a true tarpan, although many of them looked similar, such as the Hegardt or Stroebel's horse, Heck horse, and a derivation of the konik breed. Like so many horses or their ancestors, the exact classification and therefore name of the tarpan has been hotly debated over the years. It was first scientifically observed near the Russian city of Voronezh and described later by Gmelin around 1770. Later debates suggested that, strictly speaking, *E. f. ferus* might more appropriately be described as *Equus caballus* or *Equus caballus ferus*. It has generally been accepted that this may cause confusion, so *E. f. ferus* is the most commonly used name.

Unfortunately, there have been studies on only two well-preserved specimens, so relatively little is known about this horse. There have also been suggestions that two types of tarpan lived, a forest-dwelling type and a steppe-dwelling type. This has not been proven and the concept may come from the extinct predecessors of the horse, some of which were thought to have both grassland and forest subtypes. Although the forest and grassland horses may not be differing subtypes, this does not mean that the horse did not live in both environments. Forest-dwelling animals have been described in countries

such as France, Spain, Great Britain, and Sweden. The last known living individual was around 5 feet (1.5 m) tall and had shoulder and dorsal stripes with a grullo coat. The mane has also been widely disputed, with some indicating a more "wild" short, standing mane, while others indicating the falling mane that is associated with the domesticated horse. The consensus from studying historical records is of a short, falling mane.

Cave art in Spain and France shows many pictures of horses, and it is thought that these are the tarpan. Initially, the horses were hunted for meat, but progressively they were also culled so that they did not eat crops or grasslands or breed with domesticated horses, which would reduce the value of any resulting foals. The last known mare was killed in 1890 during an attempt to capture her for breeding purposes once the threat of extinction had been fully realized; the last male tarpan died in 1909.

The tarpan probably still lives on in the DNA of the domesticated horse. DNA evidence indicates that the early domesticated horses were also bred with the tarpan and therefore it contributes to the DNA variation within the horse today.

Above: *The Cave of Altamira in Spain contains many cave paintings and drawings, including the horse, from the Upper Paleolithic era. One of the three galleries within the cave is even called the "Horse's Tail."*

EQUUS FERUS PRZEWALSKII (PRZEWALSKI'S HORSE)

Despite Przewalski's horse being formally discovered by L. S. Poliakov in 1881, there remains some discussion as to whether it should be a subspecies of the wild horse as described previously (*Equus ferus przewalskii*), a species in its own right (*Equus przewalskii*), or even be included within the domestic horses as simply a subpopulation of *Equus ferus caballus*. DNA evidence shows that Przewalski's horse was not an ancestor of the modern domestic horse or vice versa. This indicates that it should not be a subpopulation of the domestic horse. The exact date at which the modern domestic horse and Przewalski's horse diverged is also a topic of debate, with recent DNA studies suggesting differing dates of between 38,000 and 72,000 years ago to 160,000 years ago.

The number of chromosomes in the domestic horse and Przewalski's horse differ, which suggests they are not a subpopulation. They can also produce healthy, reproductive offspring, which suggests that they are not an entirely different species.

Below: *Przewalski's horse (*Equus ferus przewalskii*) is a rare and endangered horse; there has been controversy on whether it is a species in its own right or a subspecies of either the domestic horse or the wild horse.*

It is thought that there were four basic types of horse: the primeval pony, the tundra horse, the steppe horse, and the proto-Arab. These basic four have now all died out or been crossbred out. The present-day horses most similar to the original four types are Exmoor, Akhal-Teke, Highland, and Percheron. The four original types played an essential role in the more than 300 recognized breeds that developed from these original horses. In the modern day, horses can be divided into types such as pony, cob, and hunter. With so many breeds to differentiate, it is often useful to indicate their role and physical features.

EQUUS FERUS CABALLUS (DOMESTICATED HORSE)

There are now about 60 million horses on the planet. This makes the horse the ninth most abundant mammal in the world, followed by the donkey at 40 million. Humans top this list followed by the cow, sheep, pig, goat, domestic cat, domestic dog, and water buffalo before you get to the horse. As we can see, the major food-providing animals are higher on the list. Surprisingly, only two countries claim to have no resident horses: Rwanda and Saint Helena.

Above: The Akhal-Teke horse from Turkmenistan going out to pasture. A group of horses can be called different names, depending on the part of the world and the type of group. Herd, team, harras, rag, stud, or string are commonly used words.

Right: The well-muscled, hardworking, intelligent Percheron, a breed of draft horse originating from France. It is still common across Europe, including France and the United Kingdom, and in the United States.

Conquering the Globe ∽

We have seen how the differing horse species and their ancestors have spread throughout the world. The proto-Arab migrated from North America and grazed in the hills between East Asia and North Africa. The primeval pony also migrated from North America but relatively later than the proto-Arab. Climatic conditions at the time were much tougher, so they became more hardy. Being well suited to colder climes, they reached the north ends of Asia and Europe. Tundra horses adapted to the subarctic tundra and Siberian swamps instead of migrating and lived through the harsh conditions. Later, we see remains of these horses in Greece and Vienna. Finally, the lean, fast, resilient steppe horse spread across Europe and Asia.

HORSE MIGRATION

The later story of the domesticated horse involves both the natural movement and migration of the horse itself but also the movement of the horse by humans. It is known that the modern horse was present in Spain more than 10,000 years ago, because the caves of Asturias show a vivid picture of an equine head. Similar paintings have been found throughout the Old World from the Stone Age. Naturally, many of these records show the existence of the horse, not if it was domesticated.

It has not been plain sailing for the horse. It was extinct in the Americas 10,000 years ago, and globally its numbers dropped to a few thousand individuals about 6,000 years ago. According to genetic, paleological, and archeological data, a Ukrainian tribe domesticated the wild horse at around 4000–3500 BCE. During the following 2,000 years, its numbers increased and the horse was seen spreading across Asia and Europe. The domesticated horse was thought to be widespread by around 3000 BCE. For example, it is known that China to Mesopotamia had the horse as far back as 2000 BCE. By 1000 BCE, we know that the horse existed in most of North Africa, Asia, and Europe.

Above: *Roman coins, such as this denarius from around 115 BCE, often depicted gods riding either on horseback or in chariots pulled by horses.*

Left: *This petroglyph/ rock carving shows two riders and their horses in Mongolia, with the Shiveet Khairkhan Mountain nearby.*

THE IMPORTANCE OF THE HORSE

Indicators of domestication include old records, such as a Roman Republic coin from 67 BCE showing a horse and its rider. Written documents also indicate the importance of the horse during everyday life, but especially during important episodes, such as wars. The Huns (of whom Attila was the famous leader) and the Mongols (ruled by Genghis Khan) used the horse during warfare. Nearly every image of the American Plains Indians shows an accompanying horse—where would the American West movies be without the horse? In times when the written word was rare, we see pictures and sculptures of the horse being used throughout the world. The first known use of the horse during sports was by the Greeks and Romans. The Trojan War (1300 BCE) is a classic example of written documents and art that tell the stories about the role of the horse down many generations. History also tells of the various burial rights afforded to the horse. With such importance being put on this useful and admired creature, a great deal of trading would have been carried out, and it still is to this day. With trading comes the movement of animals across countries and continents.

Although some domesticated horses live as feral horses, the only wild horse (*Equus ferus*) in existence today is Przewalski's horse (*Equus ferus przewalskii*). By 1966, this horse had become extinct in the wild, but it was later reintroduced into its original habitat of Mongolia. In a fascinating venture, some were also released into the Chernobyl Exclusion Zone in 1998. Here, the horses have faced very little human interaction and are reportedly increasing their population size.

Below: A worldwide map showing the evolutionary migration of the horse, from the first dawn horse Eohippus to the modern horse.

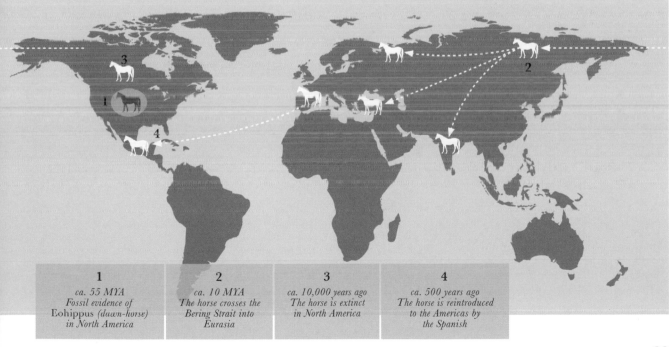

1	2	3	4
ca. 55 MYA Fossil evidence of Eohippus *(dawn-horse)* in North America	*ca. 10 MYA* The horse crosses the Bering Strait into Eurasia	*ca. 10,000 years ago* The horse is extinct in North America	*ca. 500 years ago* The horse is reintroduced to the Americas by the Spanish

CHAPTER 2

Anatomy & Biology

The Horse as a Mammal ❧

The horse is classified within the class Mammalia. There are about 5,450 species of mammals, and they are further split into three subclasses: the placental mammals, the group that the horse belongs in; the marsupials, which carry their young in a pouch, such as the kangaroo; and the egg-laying monotremes, including the platypus. The other subclasses are now extinct.

MAMMALIAN FEATURES

Mammals are vertebrates, meaning they have a backbone. Reptiles and birds are also vertebrates, but mammals differ from them in a number of ways. They have hair and three middle ear bones. They also have mammary glands, and these are used by female mammals to feed milk to their offspring.

In addition, mammals have a brain area known as the neocortex, which is part of the cerebrum. The neocortex has only recently evolved and is associated with higher-order brain functions. These include spatial reasoning, language, movement via motor commands, cognition, sleep, memory, and sensory perception. These functions include the traditional senses, such as sight, hearing, taste, smell, and touch as well as other sensory features, such as those linked to temperature, movement, pain, balance, vibration, and other stimuli. Essentially, this part of the brain works with other parts of the brain to coordinate and manage all of these functions.

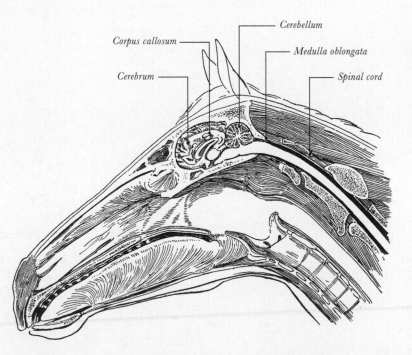

Below: The anatomy of the equine head showing the nervous system, including the brain and spinal cord. The adult horse brain weighs between 1½ and 2 pounds (680–900 g) depending on the breed, about the same size as a human child's brain.

Corpus callosum

Cerebrum

Cerebellum

Medulla oblongata

Spinal cord

The Life Cycle of the Horse ❧

The life cycle of the horse starts in the womb. Once the embryo has formed from an egg and sperm, the resulting pregnancy lasts for an average of 340 to 342 days. If a foal is born at 315 days or earlier, it is premature and considered to be at higher risk. If born before 300 days, it is unlikely the foal will survive. After birth, the horse moves through the stages of life from its first uncoordinated timid steps, frolicking through the fields, rolling, playing, maybe working, and gradually retiring to a slower paced, more gentle life in its later years.

PREGNANCY AND LACTATION

Equine pregnancy usually lasts 11 months but is often shorter in smaller breeds. The growing fetus develops inside the uterus, which is filled with amniotic fluid. The placenta is attached to nearly the whole of the uterus and connects the foal to its mother through the umbilical cord. It lets nutrients and oxygen pass from the mare to the fetus and provides for the waste material to be removed. The placenta supports the fetus until birth, and it ensures that maternal and fetal blood do not mix.

Mammals have differing types of placentae. That of the human and mouse is like a disk (discoid), while ruminants, including sheep and cows, have several connected spheres (cotyledonary). Dogs, cats, and elephants have a zonary placenta, which is a band of tissue surrounding the fetus. The pig and horse placenta looks more like a large membrane covering the entire uterus. This type of placenta is called diffuse.

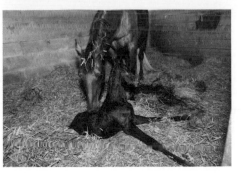

Left: *A mare shortly after birth with her foal. Foaling (giving birth) usually takes around 8 hours including expulsion of the placenta. Care of the neonate (new foal) is essential in the first few days and weeks after its birth.*

The horse has four mammary glands split into two pairs and only two teats. In the month before parturition (birth), the mare undergoes several changes in her mammary tissue. Progesterone increases the amount of mammary tissue and prolactin, cortisol, and growth hormone start lactogenesis—the production of milk. As with most mammals, the first milk, known as colostrum, is imperative for the foal. Milk is made up of lactose, lipids (fat), proteins, minerals, vitamins, and water. Colostrum also contains many immunoglobulins and antibodies to help support the immune system of the newly born offspring. These are mammalian features that are fundamental in understanding the life cycle of the horse.

In small feral mares, milk production can be up to 3 ¼ gallons (12 liters) a day, and in Thoroughbreds and large breeds it can increase to 4 ¾ gallons (18 liters), but milk production depends entirely on the needs of the foal as well as the health and ability of the mare to make milk. In comparison, dairy cows often produce up to 13 to15 gallons (50–57 liters) a day. The foal usually starts to eat other sources of food at about 8 weeks old and is gradually weaned off milk. However, some people prefer to start creep feeding at 1 week and others let lactation continue until weaning. Unmanaged mares can lactate for up to 18 months, if they are not in foal again.

THE IMPORTANCE OF SUCKLING

Once delivered, the first hurdle for the foal is to stand up and suckle. If the foal does not drink the colostrum from the mare's milk within the first 24 hours, its

body will no longer be able to absorb the immunoglobulins, so this vital opportunity for continued health will be lost. Failure to thrive at this stage can quickly lead to a sick foal. In many instances, failure to feed can be a result of foal illness.

ILLNESS

The newborn foal still has to undertake many physical changes in order to adapt to life without a placenta and outside the uterus. It is especially at risk from bacterial, viral, and parasitic infections. Hypothermia, sepsis, neonatal encephalopathy, and meconium impactions (which result in pain and swelling of the abdomen) are all problematic in the young, even causing death. Therefore carefully making sure that the foal is walking, feeding, breathing properly, not having seizures, and generally behaving appropriately is essential for its health.

Above: *Following birth, the mare must adapt to look after her young foal and the foal must adapt to the outside world. Making sure the foal feeds on milk for nutrition, liquids, and immunity is essential. A sick foal may refuse or be unable to feed naturally.*

Although the first few days are critical, a young foal will still remain susceptible to illness for its first few months. Its diet must be carefully controlled during the weaning stage, using different techniques, because the foal will be growing rapidly. Growth accelerations are seen especially within the first month and between 6 and 9 months.

THE FIRST TWO YEARS

As a yearling and 2 year old, a horse is generally at less risk from illness. At this stage, sexual maturity is observed. It is normally reached between 15 months and 2 years of age in a mare, but it can take up to 4 years or be as early as 9 months. Stallions start producing sperm at 12 to 14 months old, but they are usually more than 15 months old before they successfully breed. With this in mind, mares and fillies must be kept separately from young stallions so that inbreeding does not occur. Once sexual maturity is reached, the horse will probably undergo another growth spurt, so again it is vital to make sure of an appropriate diet to maintain bone and tissue growth.

MATURITY

Horses that mature quickly such as Thoroughbreds can be broken earlier than other breeds of horses, sometimes in their second year of life. Other horses such as warmbloods are generally broken from 3 years of age and upwards. By the age of around 3 to 5 years depending on breed, the horse will be considered a full adult. The horse will now be growing less and its diet should be carefully controlled to prevent obesity.

OLD AGE

Many people have laid claim to owning the oldest horse in the world. The *Guinness Book of World Records* has listed horses, especially ponies, in their fifties and even sixties. No doubt these are phenomenal ages, and on average ponies live longer than horses, but the equine lifespan is normally between 25 and 30 years. However, great advances in medicine have increased the lifespan of the horse in the past few decades. Making sure the horse is well fed and watered, appropriately sheltered and housed, and has adequate veterinary care and preventive medicines will all help to protect the horse and enable it to live a longer, healthier life.

Below: *An older horse resting in a field. A horse normally matures when 3 to 4 years old but lives on average of 35 to 30 years, and some have lived for more than 60 years. It really is a lifelong commitment to own a horse.*

Breeding ✑

Breeding is extremely important and entails a great deal of thought for several reasons. Throughout the years, breeding has concentrated on a number of different factors. For the Thoroughbred horse, for example, speed and agility are highly prized. For the preservation of a gene line, the look of a breed or species is paramount. For the workhorse, strength is a key feature, and for the pony, compact size is highly desired, too. In all cases, the health and, in general, the look of the offspring are of great importance. Whether investing in a multimillion dollar stud, preserving a breed, controlling feral populations, or breeding a companion pony for a child, the considerations and preparation required are essential elements to breeding science.

WHAT WILL THE HORSE BE DOING?

The future adult role or workload of the foal must always be considered. Most horses undertake some level of work. It might be as a racehorse, working the fields, pulling a trap or farming equipment, or being used as a show horse. Some owners may want to keep a horse as a companion, but it is still fulfilling a role, so its looks, conformation, size, temperament, health, and athletic ability are all vital attributes for an owner. We often talk of pedigree. This is based on a number of factors, but it is generally accepted as all of these attributes plus whether the horse is fertile, and, often in the commercial world, if it is successful in its role. In the modern world, genetic manipulation plays a large part, but even before the technologies of today some basic principles have been applied.

HOTBLOOD

The so-called hotblood horse evolved from horses living in desert conditions, and many say that its temperament is excitable, nervous, and sensitive. Typical examples inclue the Arabian horse and Thoroughbred horse.

Arab
Thoroughbred

INBREEDING AND CROSSBREEDING

In general, inbreeding (mating mares and stallions that are closely related) is generally not good for a line in the long term. In a few instances, it has been done in order to preserve rare breeds, but even in nature animals are more likely to breed with those less related to them. Stallions that are of reproductive age often leave their family unit, for example, to find unrelated mares. As with all animals, breeding with close relations increases the chance of disorders being passed on and of new mutations occurring.

Crossbreeding (mating a mare and stallion from differing breeds or even species) can bring both problems and advantages. Particular breeds are generally known for having particular characteristics and strengths, so a breeder may want to introduce these specific characteristics into another breed. Biologically speaking, crossbreeding is often fairly healthy for the offspring, because the mating horses are less likely to be related so they contain more genetic variability. Naturally, the cross must be carefully carried out. For example, a small mare may not be able to carry a foal that is abnormally large due to the larger breed of the stallion that she was crossed with, although the maternal environment tends to keep the fetus smaller in these cases.

There are many modern horse breeds, and they are usually broken down into three basic groups: the pony, the workhorse, and the sports horse.

Below: *The terms "coldblood," "warmblood," and "hotblood" do not describe the physiology of the horse. They refer to its temperament, with coldblood for placid at one end of the scale and hotblood for the more sensitive breeds at the other end.*

WARMBLOOD

The warmblood, or warm-blood, horse, such as this Hanoverian, is usually a cross between a hotblood horse and coldblood horse. Its temperament is ideal for dressage, driving, and eventing.

Hanoverian

COLDBLOOD

The coldblood horse usually originates from a cool, less arid part of the world, such as this Percheron, originally from France. It is usually a large horse used as a draft horse and workhorse.

Percheron

BREEDING RECORDS

In the wild, horses reproduce effectively, but in the case of human selection, the sire (father) and the dam (mother) are usually chosen carefully. Breeders often register young offspring in registers or studbooks, which can also be used to describe the stud males that are presently breeding. As with many profit-making industries, some of the registries or studbooks have a better reputation than others.

In general, there are two types of studbook, which are described as open or closed. The term "closed" means that no mating is allowed with outside bloodlines. An example of such a line is the national horse breed of Finland—the Finnhorse. It was first described in the thirteenth century, but the official studbook started only in 1907. The studbook was initially started because there had been a severe decline in the features and numbers of this breed, so the government called upon a concerted conservation effort. Many closed breed books concentrate on conservation of the line, either for the breed in its own right or to preserve some of the strengths within the line, for example, a particular look or a winning horse racing line known for their speed. Open studbooks also often have a variety of standards to be met but may not insist on the parents being registered. Cross-breeding may be allowed within these registers. Criteria can be strict or more open, which can lead to controversy.

The more recent type of registry is known as the Appendix. These lists often contain horses that conform to many of the criteria required in an open or closed studbook but perhaps not all. An example is the American Quarter Horse Association Appendix, where part-Thoroughbred/part-quarter horses can join this Appendix and even be transferred to the full registry based on performance.

Below left: *The paint horse is not only ornate with highly recognizable coat patterns and colors, but is also adaptable to many different equine sports, such as Western events and show jumping.*

Below: *The Friesian, dating back to around the eleventh century, usually has a black coat, but it can be chestnut, too, and can carry white marks. This classic horse is often used in movies and on television, because of both its beauty and gentle temperament.*

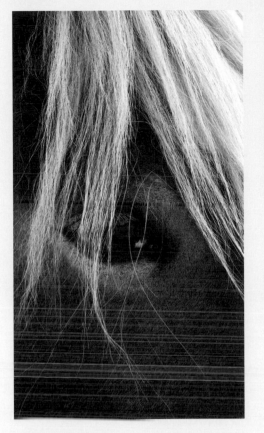

Many people have kept breeding records, from the Bedouins to fourteenth-century monks to the present day. *The General Stud Book of England* for the Thoroughbred breed is said to be the oldest of the more contemporary books, with its first volume published in 1791. It is referred to as the original "mother" studbook.

The importance of a studbook is apparent when considering the financial implications of breeding. The racehorse Frankel was valued at £100 million ($160 million at the time) and in his first year as a stud made £15 million ($24 million) in stud fees. Of his offspring from that year, 85 percent became race winners. The stud fees dwarf the £2.9 million ($4.7 million) he made in race winnings, despite these winnings being exceptional. Fusaichi Pegasus was sold for $70 million (£47.6 million), and there are many unsold horses that would command high prices. However, breeding does not guarantee a winner: The Green Monkey was sold for $16 million (£8.5 million) but raced only three times, placing third in one race, and was then retired to stud. He commanded a $5,000 (£3,100) servicing fee, but more was expected from him.

Average prices indicate that the most expensive breeds are the Arabian, Morgan, Thoroughbred, Friesian, and paint horse, though average prices vary greatly even between these breeds, from $25,000 for a paint horse to $100,000 for an Arabian.

Above left and above:
The Finnhorse, first described in the thirteenth century, is an example where conservation efforts have been used to conserve and reinstate the breed. It is now the official national horse breed of Finland Although a coldblood horse and good at drafting, it is also said to be fast.

Crossbreeding & Hybrids ∽

The breeding of Przewalski's horse has been interesting, because so few animals were available for breeding. In 1945, just 9 of the 13 remaining horses produced offspring and the population today is descended from these 9. In addition to the inbreeding observed at this point, these 9, in turn, are descended from 15 originally captured in 1900. There are also questions around the "purity," because two of the nine horses were actually hybrid horses. One was from a wild horse bred with a domestic mare and the other was a mare bred with a tarpan.

Despite the inbreeding, more than 300 Przewalski's horses existed in the wild in 2011, and in the 1990s more than 1,500 were present in captive populations worldwide. By 2014, the international studbook indicated that there were 1,988 horses in total (1,101 females, 883 males, 4 sex unknown) but it includes both captive and reintroduced animals, whether completely free roaming, living within large reserves, or housed in zoos and smaller sanctuaries. Naturally, this horse is still classified as endangered, but it was officially previously described as "extinct" until 1994, then "extinct in the wild." In 2008, it was reclassified as "critically endangered" before moving to "endangered" in 2011.

CREATING NEW SPECIES

It is interesting to understand how the horse has helped to create other species. When crossbreeding, we must also consider that offspring from species that are too different may be miscarried, be born with abnormalities, or be infertile; if infertile young are produced, it is not considered a new species. Classic examples of infertility are the mule (*Equus mulus*; the young of a male donkey and female horse) and the hinny (female donkey and male horse). The mule is said to be more hardy, strong, and have a greater lifespan than horses and to be more intelligent than donkeys. Pregnancy in a female mule is rare, with only 60 cases of birthing mares in the past 500 years. Male mules are infertile. There has been only one documented live birth in the hinny; males are sterile and it appears that most females are too. Because only one offspring has ever been documented, the hinny is not given a species name (unlike the mule) and is simply known as *Equus asinus* × *Equus caballus*. Zebroids (zebra mated with any other equine species) and more specifically the zorse (zebra stallion

Above: *A zebroid is a cross between a zebra and any other equine. Other popular names include zedonk, zenkey, zorse, zebra mule, zonkey, and zebmule. Usually a male zebra is mated with a female equine, but a zebra dam can mate with a male donkey, which produces the usually sterile and rare zebra hinny, or donkra.*

with a horse mare), zony (zebra stallion and a pony mare), and zonkey (zebra and a donkey) also exist.

The one hybrid that has been shown to be fertile is *Equus africanus* (wild African ass) × *Equus asinus* (domestic donkey), and historically ass/onager and horse/onager hybrids have been created. In 2006, a moose–horse hybrid was reported by a Canadian rancher. Veterinarians and scientists are skeptical and it is probable that the foal is just an unusual looking horse. Despite the veterinarians and scientists requesting genetic testing, it is not known whether the owner has allowed it to be done; results have not been publically released.

Some old texts speak of deer–horse crosses but these are again not proven. A few ancient texts mention the horse–human (centaur) hybrid, but most people are happier with leaving the centaurs to mythology and children's bedtime stories. One thing that these great stories do show is the relationship between humans and horses. The centaur is depicted as a strong, graceful creature in a revered position of authority, showing the respect that humans have placed on the horse for thousands of years. From the ancient Greeks and Romans to medieval monks, Shakespeare, and J. K. Rowling, this mythical creature shows an innate human fascination with the horse.

Below: *The mule is the offspring of a female horse (mare) and male donkey (jack). The mule is well known for being hardy, good natured, and patient. In 2003, it became the first hybrid animal to be cloned, and the baby clone was named Idaho Gem.*

The Mechanics of Breeding ∾

Here, we explore the reproductive systems of the horse, horse-breeding techniques, and the growth and development of the fetus. Mares become sexually mature when between one and two years old but will typically give birth to their first foal in their third year. Stallions are ready to mate between two and three years of age.

THE FEMALE REPRODUCTIVE SYSTEM

At sexual maturity, the mare will develop her estrus cycle—sometimes referred to as being "in heat," "showing," or 'in season." The female is in estrus for about 5 days and then in diestrus for around 20 days—not too dissimilar to the female human reproductive cycle. Follicles form around the eggs and, as they grow, they produce a hormone called follicle-stimulating hormone (FSH). They also produce estrogen. During this time, the mare may have changes in behavior (being more receptive to stallions, stretching her neck, and raising her upper lip), the vulva lips open and close (commonly called winking), her tail goes up, the cervix relaxes, and she may urinate more often. On the fifth day of estrus, a follicle ruptures due to the hormone called luteinizing hormone.

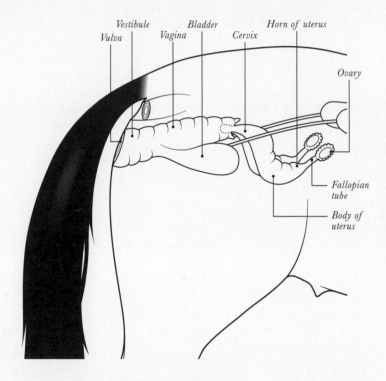

Vestibule Bladder Horn of uterus
Vulva Vagina Cervix Ovary
Fallopian tube
Body of uterus

Above: The female reproductive system is completely developed in most horses by the age of one or two years.

The egg is released and a follicle becomes the yellow body (corpus luteum) that releases the hormone progesterone. The egg travels along the Fallopian tube and can become fertilized. If the egg is fertile, it will travel to the uterus. This cycle will normally happen across three to four months a year, but it does vary and is particularly responsive to daylight length; ovulation is more likely to occur in the months where there is a lot of daylight.

THE MALE REPRODUCTIVE SYSTEM

Sperm is needed to fertilize an egg. Sperm production is affected by light, so it is more efficient from April until August. During this time, a stallion produces more sperm (up to eight million cells per day), he will become more receptive to sexual stimulation (an average time of just two minutes in the summer compared to ten minutes in winter), and will ejaculate in a fewer number of mounts (an average of one time in summer compared to two-and-a-half times in winter).

Some common problems are observed in stallions. Overuse can be a problem and different stallions will have differing limitations on uses per day. A condition called hemospermia, where blood is observed in the sperm, can result in sterility. Cryptorchidism is the term used to describe the situation when one or both of the testes do not descend. If only one testis is affected, the male may be fertile but he should be castrated, because this disorder is inherited. Normal illness can affect the performance and fertility of a male, and a number of reasons, including illness and youth, can prevent a male from ejaculating. Age, general health, and management are always important issues in both the male and female. If a male has Klebsiella and/or B-hemolytic streptococci, the female will be less likely to conceive. Males infected with equine viral arteritis (EVA) must be used carefully, because a female who is infected after conception can abort.

For these reasons, the stallion used is important and his health and well-being are essential. Although stallions will cover naturally, in more managed or expensive horses, a number of techniques such as artificial insemination are used to make sure that the male and female are not damaged in the mating process.

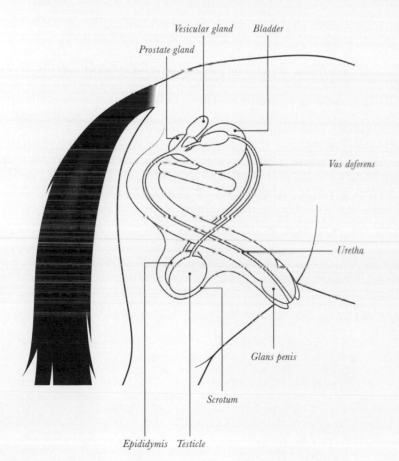

Left: *The male reproductive system is normally mature by about 18 months, but many breeders will wait until the colt is three years old to breed from him.*

Prostate gland

Vesicular gland

Bladder

Vas deferens

Uretha

Glans penis

Scrotum

Epididymis

Testicle

ASSISTED REPRODUCTION

Many adaptations can be made to the natural estrus cycle in the mare. For example, veterinarians can control the cycle using hormones and even induce estrus during the winter months. Hormone therapy is often coupled with light therapy to induce this change. In the 1960s and 1970s, studies started to show the positive pregnancy rates for artificial insemination (AI) in comparison to a natural service. The first formal and informal trials, both experimental and commercial, were held well before this period and, of course, technology continues to develop. AI has many benefits: It reduces the need for horses to travel extensively for breeding purposes; it can reduce disease contraction, if tested appropriately; a stallion can be used even if abroad; and precollected sperm can be used for years after a stallion has died or a gelding is castrated. It also has some cons: Success rates are not always as high as natural conception; some stud books do not allow AI; and disease can arrive into countries and areas where it is not present.

Many owners and breeders still use fresh semen or a natural service, while others are happy to use frozen or chilled sperm. In 2017, research indicated that a new technique may preserve sperm at room temperature for weeks instead of two or three days. This will be tested in more detail in full trials. Similar advances are being made across the board from fresh to cryopreserved sperm.

The use of embryo transfer is also on the rise, thereby allowing the egg and sperm from the male and female to be put together under controlled conditions and the embryo placed into a surrogate mare.

Below: *Artificial insemination and embryo transfer are increasingly popular in many species, including the horse. It enables sires from different countries, for example, to inseminate and enables surrogacy.*

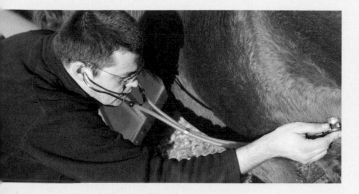
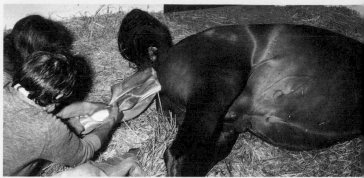

This technique is frequently used to breed more offspring from outstanding mares, move embryos to mares farther away, or preserve embryos for future implantation. Costing many thousands of dollars per implanted embryo, this method is not cheap, so it is usually reserved for foals expected to sell at a high price or for valuable conservation work.

PREGNANCY

Gestation in horses usually lasts 320 to 370 days, with an average of 342 days; it is generally shorter for smaller breeds and longer for larger breeds. Pregnancy can be identified through a number of methods. An owner or breeder can look for behavior changes and physical shape changes in a potentially pregnant mare. A skilled practitioner can identify pregnancy by a rectal examination at 20 to 30 days, but it is more accurate between 40 and 50 days. Transrectal ultrasound can also be used and is excellent for identifying twins. It is frequently used between 12 and 15 days and is useful in helping to ensure a single pregnancy, because action can be taken for mares carrying twins.

Common blood tests include ones that test for progesterone (days 18 to 25) and pregnant mare serum gonadotrophin (PMSG). The latter is called the equine chorionic gonadotropin (eCG) test or mare immunological pregnancy test (MIP), is about 95 percent accurate, and is used at days 40 to 120. Estrogens, especially estrone sulfate can be tested from about 110 days to term and is about 99 percent accurate. PMSG should not be used in donkeys, and these tests can be inaccurate outside of their suggested dates. Not all of the tests are always accurate; this is especially true in the 90-to-120-day period. Estrogens in the urine can also be tested. Urine is relatively easy to obtain and the test is most accurate from 100 days.

Twins are problematic in equids. About 80 percent will naturally miscarry and, even if carried to term, complications in the mare and foals are common and can be life threatening. The equine placenta covers the entire uterus, unlike many other mammals. In humans, for example, there can be two placentae or a larger placenta, but this is not possible in a horse, so the growing foals do not receive sufficient nutrients and gases. Twin foals have the same combined weight as a single foal, and they do not usually gain size after birth. With the risks to mare and foals, the decision to terminate one foal early on can provide the safest outcome.

Above left: *Veterinary assistance can be required at many stages of breeding, ranging from prebreeding checks, pregnancy and twin confirmation, evaluating mare and fetus development, and for assisted foaling in addition to general health checkups.*

Above: *Usually only one foal is carried to term in a pregnancy, but, on rare occasions, twins are born. Foaling often does not require medical assistance from humans, but it can be required.*

Anatomy of a Horse ∽

The anatomy of the horse depends on its genetic makeup. In itself, the horse differs from all other mammals. This might be in small ways to closer relatives, such as the zebra and other near relations, and in larger ways in comparison to other species, such as the human or dog. The genes dictate how the body develops from the single cell to a complex organism. Genes control bone and muscle growth, where the blood vessels run throughout the body, how the heart beats, and even which hormones are released.

Right: *Common external anatomical regions. When describing equine anatomy, it is often useful to use regions or parts of the body instead of the full anatomical names for each bone or muscle. This diagram shows some of the commonly used terms.*

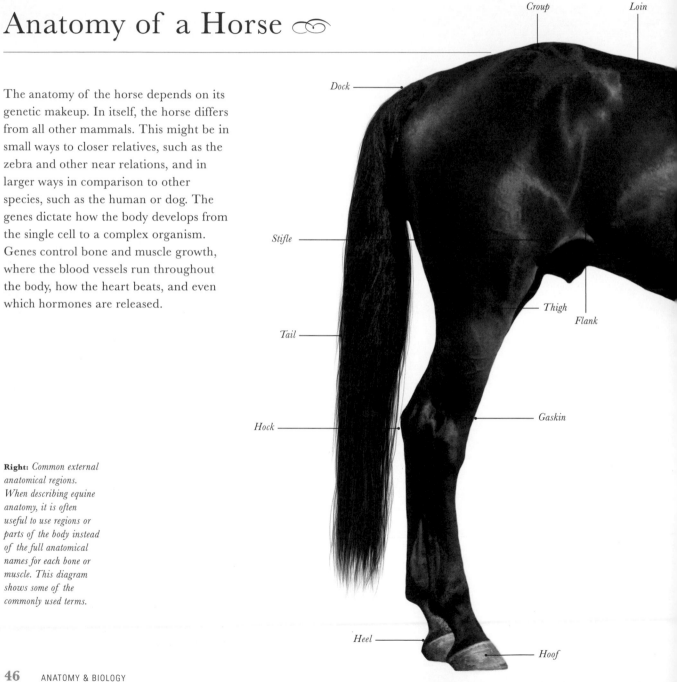

Croup

Loin

Dock

Stifle

Thigh

Flank

Tail

Hock

Gaskin

Heel

Hoof

Crest

Poll

Forehead

Withers

Back

Chin groove

Throat

Muzzle

Shoulder

Breast

Elbow

Forearm

Knee

Cannon

Fetlock

Pastern

Coronet band

The environment can affect a horse's health. If the horse is malnourished, it cannot take in the nutrients and minerals required for strong bones and muscle development. Likewise, if the horse is well exercised, the muscles will get larger. A horse in cold climates will probably have more hair and be stockier in build, so it will lose less body heat than those in arid conditions. Injuries and diseases also play a large role in the physiology of the animal. Therefore the health and how the horse functions is based upon a careful balance of breeding (genetics) and the environment in which it lives, including food, drink, climate, exercise, chemical exposure, disease, and injury.

The Skeleton & Body ❧

The skeleton has many functions. It provides the main basis for locomotion. The muscles and other body tissues attach to the skeleton, so it provides the basic outline and support for the body, too. It also protects vital organs from damage, for example, protecting the brain within the skull and the spinal cord within the vertebrae. Many bones also produce blood cells, both red and white. These are essential for transporting oxygen around the body and as a defense mechanism against foreign bodies, such as bacteria. Bones also store minerals, including calcium and phosphorus.

THE SKELETON

As the embryo grows, the skeleton is initially made from the more supple cartilage, but most of it is gradually replaced by bone. This occurs throughout gestation and bone is then remodeled according to the environment until the horse's death. Adult horses maintain cartilage in their bodies as a connective tissue for many uses. Examples of places where it persists are in the ribs to enable the rib cage to expand, in the ears to let the ears fold, and in the trachea to help keep the airway open. The skeleton is normally split into two regions, either appendicular (fore- and hindlimbs) or axial (all other structures/bones). An adult horse has 205 to 207 bones: 34 in the skull, 40 to 42 in the forelimbs, 40 in the hind limbs, 37 bones make up the ribs and sternum, and the spinal vertebrae are made up of 7 cervical, 18 thoracic, 6 lumbar vertebrae, 5 sacral (these are fused), and around 18 coccygeal bones (also called caudal).

Below: *Horses normally have around 205 bones in their body. They are vital for providing support, protection, producing blood cells, and storing minerals.*

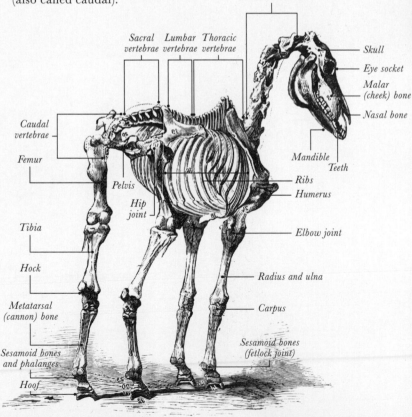

Cervical vertebrae

Sacral vertebrae Lumbar vertebrae Thoracic vertebrae

Skull

Eye socket

Malar (cheek) bone

Nasal bone

Caudal vertebrae

Femur

Mandible

Teeth

Pelvis

Ribs

Humerus

Hip joint

Tibia

Elbow joint

Hock

Radius and ulna

Metatarsal (cannon) bone

Carpus

Sesamoid bones (fetlock joint)

Sesamoid bones and phalanges

Hoof

MUSCLES, TENDONS, AND LIGAMENTS

Working alongside the bones and cartilage are the muscles, tendons, and ligaments. Muscles are subdivided into three types: skeletal, cardiac, and smooth. There are around 700 skeletal muscles in the equine body, and they are primarily used to support the skeleton, to enable movement and locomotion, and to help with the act of shivering when the horse needs to warm up. Cardiac muscle is found only in the heart, as suggested by the name. The horse has no conscious control over cardiac muscle, because it is controlled from the central nervous system to ensure that the heart contracts. Smooth muscle is also controlled by the same system and is located in a number of regions throughout the body. It helps contraction of blood vessels, the bladder and intestines, and plays major roles in the digestive and reproductive systems.

The tendons work with the muscles in that they essentially attach the muscle to either another muscle or to a bone or cartilage. Tendons are fairly elastic, so they can give a good range of movement.

Ligaments can be elastic or inelastic and are generally tougher and less flexible than tendons. This reflects the differing roles that ligaments have. They hold bones together or even tie bones to each other, generally supporting tissues and bones, and they can help "guide" tendons by wrapping around them, thus maintaining direction and giving support.

Together, the muscles, tendons, and ligaments support and move the skeletal system. They also control the types of movement that each joint or bone can do.

Right: *Each horse and breed may differ slightly from each other, but the overall musculature is similar in most horses.*

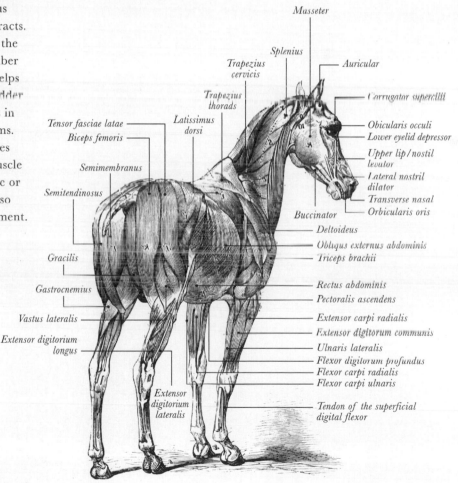

Masseter
Splenius
Trapezius cervicis
Auricular
Trapezius thorads
Corrugator supercilii
Latissimus dorsi
Obicularis occuli
Lower eyelid depressor
Tensor fasciae latae
Upper lip / nostil levator
Biceps femoris
Lateral nostril dilator
Semimembranus
Transverse nasal
Semitendinosus
Orbicularis oris
Buccinator
Deltoideus
Obliqus externus abdominis
Triceps brachii
Gracilis
Rectus abdominis
Pectoralis ascendens
Gastrocnemius
Extensor carpi radialis
Vastus lateralis
Extensor digitorum communis
Extensor digitorium longus
Ulnaris lateralis
Flexor digitorum profundus
Flexor carpi radialis
Flexor carpi ulnaris
Extensor digitorium lateralis
Tendon of the superficial digital flexor

VITAL SYSTEMS

Within the body there are many vital systems. The reproductive system has already been looked at in detail. The sensory systems are looked at later in this chapter, as is the digestive system alongside the urinary system. In addition to these are the cardiovascular, lymphatic, and respiratory systems, which control the heart, blood vessels, and breathing.

Respiratory System

The respiratory system starts with the mouth, nostrils, and nasal passages. Humans can breathe through both their mouth and nose, but the horse can breathe only through its nose. On average, an adult takes between 8 and 16 breaths per minute when resting, but the amount of breaths is increased in younger horses. When exercising, the body needs more oxygen, so this rate can increase to more than 120 per minute.

Air passes into the nasal passages, into the pharynx, and past the epiglottis, which opens when the horse inhales but closes off when the horse eats. This enables air to pass into the top of the trachea but ensures that food and water do not. The air travels past the voice box (larynx) and into the trachea (windpipe). From here, the air travels into the lungs through smaller tubes called bronchi and bronchioles until it reaches the alveoli. These contain air sacs surrounded by many capillaries. Oxygen diffuses into the bloodstream and is taken to the heart, which pumps it around the body. When the blood returns to the heart, it is pumped back to the lungs, where the waste gas carbon dioxide diffuses back into the air sacs and is breathed out.

When the horse breathes in, space must be created for the air entering the lungs, so the rib cage expands and the diaphragm flattens. When they resume their positions, the air is forced to return out of the lungs and out through the trachea as the pressure within the chest increases. Respiratory infections are common in horses and decrease the effectiveness of the system. In addition, adaptations must be made during exercise. The nostrils, nasopharynx, and larynx dilate as the horse pushes its head forward, which helps air enter and leave the system.

Below: *The anatomy for the respiratory system in the horse differs from many other species and can be prone to infection.*

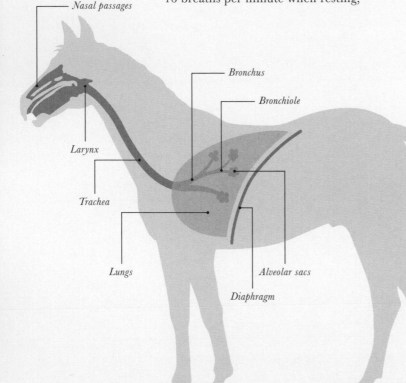

Nasal passages

Larynx

Trachea

Lungs

Bronchus

Bronchiole

Alveolar sacs

Diaphragm

Cardiovascular system

As we can see from the respiratory system, the cardiovascular system is closely linked with gaseous exchange. An exercising horse may have 260 heart beats per minute and a fit horse can have a resting heart rate of just 26 beats per minute. The heart makes sure that oxygen is delivered to every cell in the body and carbon dioxide is removed. A network of blood vessels is required for this process. Oxygenated blood travels from lungs to the heart, which then pumps the blood into arteries, followed by arterioles and then smaller capillaries. Along the way oxygen diffuses into the tissues and cells. Deoxygenated blood travels to the ends of the capillaries and into venules, then through veins and back to the lungs to deposit the carbon dioxide and pick up more oxygen.

As well as transporting gases, the blood also transports nutrients from food, water, minerals, and white blood cells, which are an essential part of the immunity defense mechanism. Lactic acid is taken from the tissues to the liver to be broken down, and urea travels from the liver to the kidneys to be processed and disposed of in the urinary system. Hormones and glucose also travel around the body. Interestingly, heat travels from the more active parts of the horse to other areas. The blood is constantly checked by a number of processes to maintain homeostasis. The horse is subconsciously checking hydration, bacterial and viral infection levels, temperature, tissue damage, hormone levels, nutritional status, oxygen levels, and many other factors. If these are not appropriate, the brain sends signals to the body to counteract the problem detected.

The lymphatic system often shadows the vascular system and has its own lymphatic vessels along with a system of lymph glands. This system helps to remove excess fluids, toxins, and bacteria, return useful resources, such as proteins, back into the blood system, and transport fats around the body. This system is largely ignored by both horses and their owners until infection sets in. At this point, the glands can often swell as a response to fight infection and the round masses can be felt under the skin.

Below: *The equine cardiovascular system undergoes many changes from embryo to newborn foal, and even during pregnancy.*

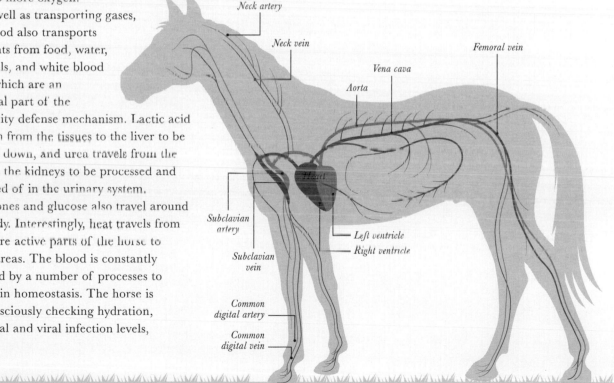

Neck artery

Neck vein

Femoral vein

Vena cava

Aorta

Subclavian artery

Subclavian vein

Heart

Left ventricle

Right ventricle

Common digital artery

Common digital vein

The Skull, Teeth & Jaw ⌒

Because the horse is an obligate herbivore (it is unable to eat meat), the skull, jaws, and teeth have evolved to be suitable for the food that the horse, and equine ancestors, can digest. As the horse evolved into a grazing animal, the jaws and muscles became strong and large in order to masticate (chew). There is, however, a balance that must be achieved between having large jaw bones and teeth and the weight of the skull in general. The horse skull therefore contains a number of sinuses (air-filled spaces) to achieve a balance between weight and strength. Some of these sinuses also play a vital role in dentition. For example, the superior maxillary sinus houses the roots of the last three molars. Infections can spread into these sinuses, which can then become infected, too. The sinuses are also important in providing space for the large teeth, protecting the brain, letting blood vessels and nerves pass through to differing parts of the head, and in positioning of structures, such as the eyes.

The lips and teeth are designed to grasp and cut food. Several glands then produce saliva in order to lubricate and wet the food so that it is able to slide down the digestive tract without causing damage. In addition, the jaw is able to move in a lateral movement that helps with grinding.

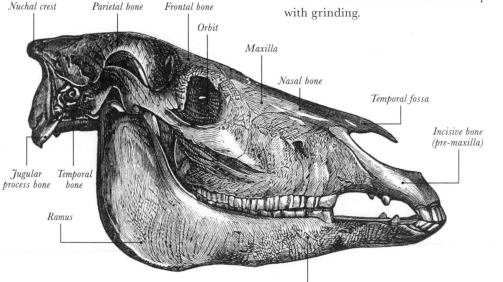

Nuchal crest   Parietal bone   Frontal bone

Orbit

Maxilla

Nasal bone

Temporal fossa

Incisive bone
(pre-maxilla)

Jugular
process bone   Temporal
bone

Ramus

Mandible

Left: *The horse skull containing the large mandible and ramus for chewing and the multiple air-filled spaces (mainly sinuses) within the skull to keep it light in weight.*

Right: *Dentition is essential for the horse to bite and chew food. Although teeth stop growing about seven years after they erupt, they can be pushed farther out of the jaw over the many years that follow.*

TEETH

The incisors predominantly bite the food whereas the molars at the back chew the food. Most of the teeth are tall and continue to grow, known as hypsodont. The teeth stop growing about seven years after they erupt. They can look as if they are continuing to grow, but they are simply being pushed out of the jaw and into the mouth at a rate of an eighth of an inch (2–3 mm) a year. The visible part of the tooth is often called the clinical crown and the longer part hidden within the jaw is termed the body of the tooth.

The premolars have evolved over time to look like molar teeth; this process is termed molarization. The enamel forms several ridges and folds, which further increase the surface area of each tooth. These features help with the chewing and grinding required by the horse. In the horse, the premolars and molars are often referred to as the cheek teeth. The cheek teeth in the lower jaw are more closely arranged together than those in the upper jaw, so the lower ones are also more effective at grinding.

The remaining teeth observed in the horse are canines and incisors. The canine teeth are usually present in only

male horses. Unusually, the first premolar (wolf tooth) does not always erupt in the horse. These two teeth, the canine and the first premolar, are completely grown when they do erupt and do not continue to grow (known as brachydont).

Above: *With lips and teeth that work in conjunction to grasp and cut food, the horse jaw facilitates the mastication of food with its lateral grinding motion*

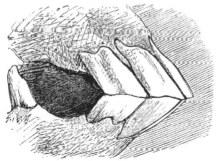

Twelve Years

Twenty Years

Thirty Years

DENTITION

The adult horse usually has 12 incisors (6 upper and 6 lower), 4 canines (2 upper and 2 lower, but usually not present in the female), 12 to 14 premolars (6 or 8 upper and 6 lower) and 12 molars (6 upper and 6 lower). The female can have 36 to 40 teeth, depending on the number of canines erupting, and the male between 40 and 42 teeth, depending on how many of the two first premolar teeth are present. The young horse has 24 milk teeth, also called deciduous teeth. These consist of 12 incisors and 12 premolars.

It is normal for a newborn foal to have a dental check. Some teeth can be present at birth but the third incisors, for example, often take three to nine months to come through. Likewise the adult teeth erupt at different times. These range from five to six months for the first premolars to five years for the canines. In general, the adult dentition is observed by five years of age, often called a "full mouth."

When a horse is around two to three years old, owners often check for wolf teeth and remove them if present to prevent later problems and to assist with fitting a bit. Some owners ask their equine dentist to smooth and round the edges of the first premolars once the horse is being broken in. This is thought to create a "bit seat," where the edges of the teeth cannot so easily rub on the inside of the cheeks. There is no evidence yet to suggest this improves performance.

The cup is usually absent from the teeth in horses 12 years old and older, so it is often easy to age horses by their teeth when between 6 and 12 years old, but it is more complex thereafter. At 9 to 10 years old, a Galvayne's groove may be present,

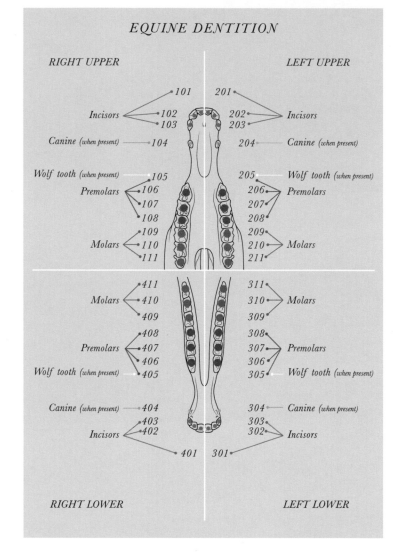

which undergoes anatomical changes as the horse ages. At 15 years old, the groove reaches halfway down the incisor tooth and, by 20 years, extends down the entire tooth. By 25 years, the groove is absent from the upper half of the tooth and, by 30 years, it has entirely disappeared. This helps to age an older horse. Aging using teeth should be a rough guide, because the type of food and environment that the horse lives in can affect dentition.

Above: *Different teeth can be present throughout the development of the horse, from a newborn foal to an older animal. Because dental problems can cause serious side effects, regular checkups are advised.*

DENTAL PROBLEMS

As with most species, dental problems are frequently observed and can show differing symptoms. Common features for a grazing horse are blood loss from the mouth, weight loss, rough hair, unpleasant odor from the mouth and/or nostrils, loss of food when chewing and inefficient chewing, producing too much saliva, head tilting, sensitive cheeks, and pain when drinking. When the horse is under saddle, it may also show signs of a dental problem by head throwing or bracing against or sensitivity to the bit. The horse's performance may be affected and it may refuse commands or be less willing to collect, may show poor head carriage, and there may be tail swishing.

Common dental problems to watch for include dental caps, molar hook, parrot mouth, wave mouth, and enamel points. Dental caps are when the milk teeth are still in place after the deciduous teeth have erupted. They must be removed,

because they can prevent normal eruption and displace or impact upon the adult teeth and can even cause infection or laceration of the soft tissues, such as the tongue, cheek, and gums. Molar hook occurs when the upper and lower jaws are not aligned properly, therefore the second molar keeps growing and can form a hook. The last molar on the lower jaw can also be affected. A "parrot mouth" is seen if the upper jaw is longer than the lower jaw. It is a congenital problem and hooks can develop on the molar teeth, necessitating their removal, and incisors can continuously grow, so may need shortening. In wave mouth, which is often seen with parrot mouth, the molars grow longer than others, resulting in differing heights despite wear to the teeth. Enamel points can develop on molars as they erupt throughout the horse's lifetime. These can become sharp and interfere with the sides of the cheeks and tongue causing ulcers and lacerations.

Below: One of the methods used for aging a horse is by looking at the development of the Galvayne's groove, which gradually regresses as the horse ages.

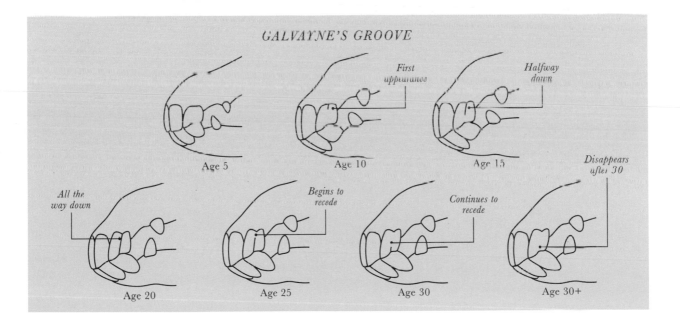

GALVAYNE'S GROOVE

First appearance

Halfway down

Age 5   Age 10   Age 15

Disappears after 30

All the way down

Begins to recede

Continues to recede

Age 20   Age 25   Age 30   Age 30+

Eating & the Digestive System ❦

The digestive system of the horse consists of the foregut and the hindgut. The foregut includes the mouth, pharynx, esophagus, stomach, and small intestines; the hindgut includes the cecum, large and small colons, rectum, and anus. The horse must ingest enough nutrients and water to make sure that the body is appropriately sustained. Once food is within the mouth, bicarbonate is released to help buffer the acid environment of the stomach. The saliva a horse produces also contains small amounts of amylase, which helps in the first stages of digestion.

The stomach works optimally when it is up to half full. Food stays in the stomach for 45–120 minutes, and a valve prevents food from being passed back into the esophagus, which means that horses are rarely able to bring food back up. This is problematic if the horse eats too much or eats something toxic or poisonous.

DIGESTION

The food passes through the throat and into the esophagus, which is 4 to 5 feet (1.2–1.5 m) in length. Food digestion then really begins in the stomach, where gastric juices, including enzymes, and hydrochloric acid start to break down the food. The stomach can be small when empty but can hold up to 5 gallons (18 liters) in a large horse when full.

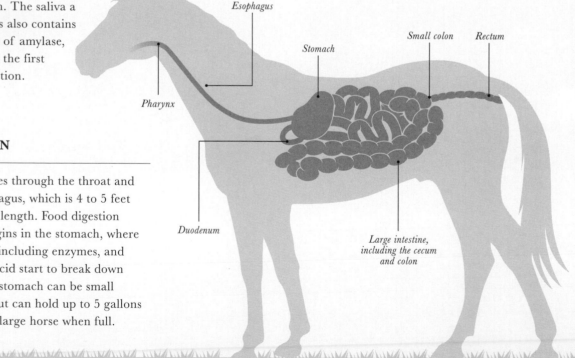

Esophagus

Stomach

Small colon

Rectum

Pharynx

Duodenum

Large intestine, including the cecum and colon

Left: The digestive system is an essential part of ensuring that nutrients and liquids pass into the equine blood vessels to feed every cell in the body. Colic is also a regular problem in horses, because they have a relatively long digestive system.

The intestines do the majority of the food breakdown, and it is here that nutrients pass from the digestive system into the blood system and are transported around the body. The pancreas helps to control blood sugar levels by releasing a range of hormones and enzymes. The liver produces bile, which is essential for digesting fats. Unlike humans and many other species, the horse does not have a gallbladder so the bile is produced continuously. The food travels through the 50-to-70-foot (15–21 m)-long small intestine, into the cecum, a 4-foot (1.2-m)-long pouch and the first part of the large intestine. It continues its journey through the next 24-foot (7-m) length of the large intestine, where bacteria help to break down fibers in the food, digestion of the food continues, and water and vitamins continue to be absorbed into the bloodstream. The food must travel through about 26 feet (8 m) of colon before reaching the rectum and passing through the anus. The horse has evolved to be able to eat plenty of roughage and fiber while also dealing with the fluids found in fresh grasses and vegetation.

GRAZING

Historically and in the wild, horses graze, eating frequently throughout the day. Many working and companion horses are fed less frequently each day; however, where possible, continuous provision of fiber is better for the digestive system. Horses are sensitive to changes to food and to feeding schedules, and as a result they are susceptible to colic due to the feeding regimes practiced by humans. Food must be carefully checked to make sure it does not contain mold or inappropriate bacteria, which can quickly make horses ill.

Horses need up to 12 gallons (45 liters) of water a day (for a 1,000-pound/450-kg horse), but it will naturally increase in hot weather and when working or exercising. It can also increase significantly more for a pregnant or lactating mare.

Systems & Senses ⮜

Like most mammals, a horses's main senses are hearing, smell, sight, taste, and touch. In addition to these, there are many other parts of the sensory system that are involved in coordinating the body and in helping the horse to respond to its environment.

BODY SYSTEMS

The senses, in general, are controlled by the nervous system. The two main parts of the nervous system are the central nervous system, which includes the brain and spinal cord, and the peripheral nervous system, which includes all the other nerves throughout the body. The peripheral system connects to the spinal cord and it sends and receives information to and from the brain. The main features are neurons, also called nerve cells. Impulses are transmitted along these neurons so that the body is able to transmit and receive information.

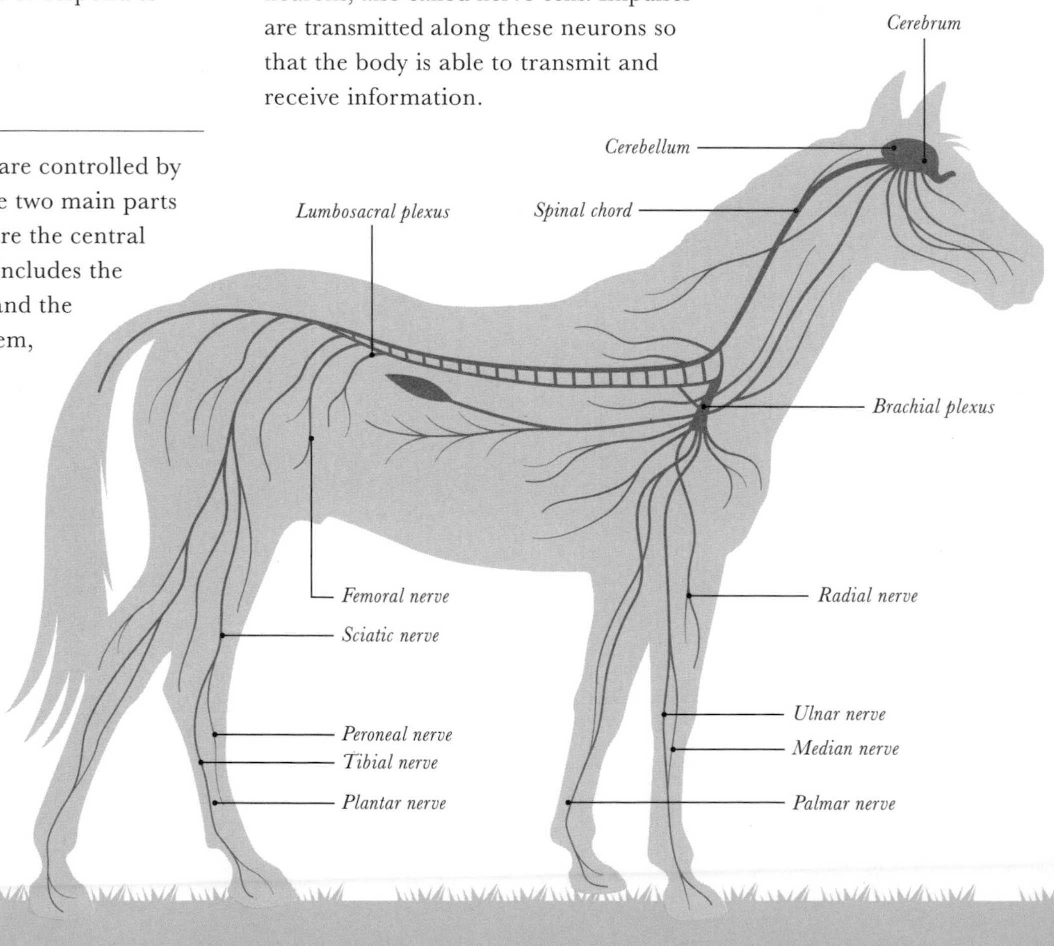

Right: *The horse's sensory systems has two major parts. The first is the central nervous system, including the brain and spinal cord. The second is the peripheral system, which includes all the other nerves that run throughout the body. Neuron cells are the main way in which this system transmits signals.*

Cerebrum

Cerebellum

Spinal chord

Lumbosacral plexus

Brachial plexus

Femoral nerve

Sciatic nerve

Radial nerve

Peroneal nerve

Tibial nerve

Ulnar nerve

Median nerve

Plantar nerve

Palmar nerve

Alert and interested   *Sleepy or unwell*   *Relaxed, listening to rider*   *Aggressive*

The endocrine system controls hormone release; these hormones regulate the activities of cells, affecting functions such as growth, metabolism, blood pressure, and reproduction.

The vestibular system, which is in the ear, is made up of the semicircular canals and the otoliths. The former detect rotational movements and the latter linear accelerations; together these provide a horse with a sense of balance and spatial orientation.

Proprioception is the sense of movement and position. It also helps the brain to work out where each joint and limb is at any given time, thus making sure that movement is smooth and coordinated. Proprioception also helps with position of the inner organs and tissues—something we do not generally think of but that is essential for the body. Proprioception both reacts to and causes conscious and unconscious processes, so the horse can make an adjustment to movement by thinking about it, or the system may work without the horse knowing.

Combined, proprioception and the vestibular system, along with the information a horse receives through its eyes, allow it to coordinate and control its movement and speed.

Nociception helps to detect danger in the forms of heat and cold, cuts and abrasions to the body, and chemicals that may cause damage or burning.

Thermoception also plays a role in how heat or cold can be detected. Receptors throughout the body ensure that the central nervous system enables the body to respond appropriately.

HEARING

Audition is highly developed in horses. Compared with humans, for example, they hear higher and softer tones. Horses may react to stimuli that we cannot hear, and so they can seemingly behave erratically or unexpectedly at times. Males may also react more readily to sound, not necessarily because their hearing is better, but because they have evolved to take on a more protective role over the herd. They may want to alert the "herd" to danger, even if that danger is an owner or empty field. It has been suggested that horses hear high-frequency sounds because they rely on using their outer ear to determine where the noise is coming from, and not because they utilize high-frequency communication. Ears are also important, because horses use them to indicate moods and behavior.

Above: *Horses have well developed hearing, but they also use their ears to portray emotions and give visual signals to other horses and, of course, humans.*

The structure of the ear

As in humans, the horse's ear comprises three main parts: the outer ear, the middle ear, and the inner ear.

The outer ear is surrounded by several muscles that help to maneuver the ear around, and it contains cartilage to maintain the ear's shape and to also help move it. There are no bones in this section, which allows for the ear to be more flexible, a feature that is essential for movement and also for letting the ear bend, for example while going through tall bushes. The outer ear helps to direct sound into the middle part of the ear.

The eardrum sits between the middle and outer ear. The sound creates vibrations that hit the eardrum. The middle ear contains the tiny auditory ossicle bones that transmit the sound vibrations from the outer ear into the inner ear. The middle ear also has a connection to the Eustachian Tube, which runs from the rear of the nose to allow air to enter.

The inner ear is protected within the skull and contains the cochlea and semicircular canals. The canals detect not only movement and angle of the head, but also contain tiny hairs and crystals called otoliths. The movement of these crystals helps to send nerve impulses to the brain. Sound vibrations pass through the cochlea and to the nerves where the signal of "sound" is passed to the brain. Normal gradual hearing loss begins at around five years of age. Usually, hearing is affected in the higher frequencies; because we do not hear these frequencies, we won't realise a horse is no longer responding to them. Their keen sense of hearing is why horses prefer to be approached from the front and why they may spook at unknown noises.

Ears should always be checked for scratches, infections, ticks, and mites and be regularly cleaned if dirt enters the ear. This will help preserve hearing and prevent pain from infections, scratches, or other accidental damage.

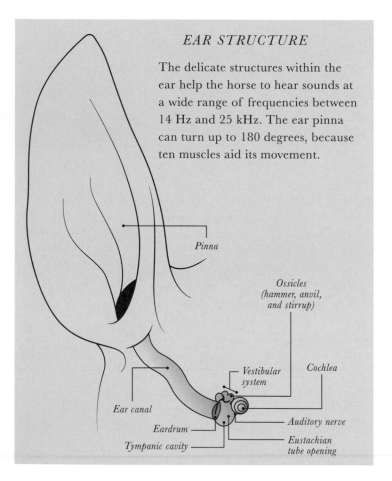

EAR STRUCTURE

The delicate structures within the ear help the horse to hear sounds at a wide range of frequencies between 14 Hz and 25 kHz. The ear pinna can turn up to 180 degrees, because ten muscles aid its movement.

Pinna

Ossicles
(hammer, anvil,
and stirrup)

Vestibular
system

Cochlea

Ear canal

Eardrum

Auditory nerve

Tympanic cavity

Eustachian
tube opening

Left: *The horse ear comprises three main parts: the inner ear (the cochlea and semicircular canals), the middle ear (the ossicle bones and the Eustachian tube), and the outer ear (the pinna, ear canal, and eardrum).*

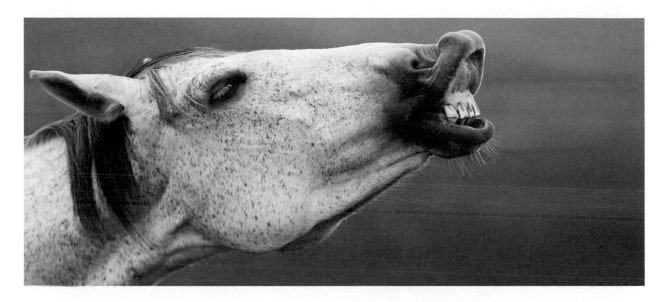

OLFACTION

The sense of smell or olfaction is essential in finding water and food, avoiding danger, and, for the males, detecting females that are "in season." Horses also know each other by smell and, as with many other mammals, it is one of the factors used to assist with mother–foal bonding.

The nostrils lead into the nostril chambers. They can dilate in order to increase the amount of air entering the chambers. The false nostrils help to filter the air and prevent infection by catching dust and bacteria on the tiny hairs that line the skin. As the air travels in, it passes the turbinate bones, which are covered by membranes that secrete mucus. The many millions of olfactory cells are located on this membrane and without them a horse would not have a sense of smell. These sensory cells are stimulated and the nerves send impulses to one of the two olfactory bulbs, which are found on the front of the cerebrum (the principal part of the brain. The left nostril sends impulses to the left olfactory bulb, likewise on the right-hand side.

In horses, there is a secondary olfactory system that involves the vomeronasal organs. These are under the nasal cavity, and a duct called the nasopalatine duct creates a link to them. When a strong odor stimulates the sensory cells, a signal is sent straight to the brain. It is thought that this second system mainly deals with pheromones and chemical signals from other horses and perhaps other species. The horse may change its behavior and hormone levels may be altered, for example, following stimulation from an in-season mare. We may observe the horse using this sense of smell when it curls its upper lip and lifts its head up. This action helps trap air and enhances the ability to smell. The horse laugh is really a horse smelling life. The vomeronasal system does not exist in humans. Smell is often linked to taste as the two combined can, for example, send strong signals and assist in food choices.

Above: *Similar to hearing, a horse usually has a higher olfactory sensitivity than most humans. Your horse can smell you from around 100 paces away.*

Below: *Touch and smell are vital requirements for both domesticated and feral horses.*

Right top: *Horses have dichromatic, two-color vision that helps them see yellow and blue but not red.*

Right bottom: *Because the horse is a prey animal, the eyes are situated on the sides of its head. "Blinkers" or "blinders" are often used to calm horses or keep them from being distracted.*

TASTE

Horses usually prefer sweet and salty tastes to bitter or sour foods though this can differ depending on the sex of the horse. Their sense of taste (gustation) is well developed and helps to protect them from eating poisonous plants—they will avoid bitter-tasting plants.

The majority of the 25,000 equine taste buds are found on the tongue but some are also present on the palate of the mouth and in the throat. Once the food is taken into the mouth, saliva covers it and the chemicals are dissolved. The chemicals can then pass through the taste buds to the taste receptors cells, which are located on structures called papillae. Different tastes are detected based upon location and type of the taste bud, and the brain determines the taste of the food from the nerve signal.

This food checking system is much more important in the horse, which cannot vomit, than in an animal that can expel poisonous food. The fondness for sweet foods is perhaps not as useful for the domesticated horse with plenty of nutritious feed available, but in the wild horse the preference for sweet, calorie-rich foods may have helped them to gain weight. Naturally, the sense of touch (texture) and smell of the food will also assist the horse in determining whether it is safe to eat or not.

TOUCH

The somatosensory system is responsible for feeling touch. This sensory system can feel changes both inside the body and at the surface of the body. This is where there are differing receptors, such as the thermoreceptors (temperature), mechanoreceptors (mechanical pressure or distortion), chemoreceptors (chemicals), and nociceptors (dangerous stimuli, such as extreme heat/cold, pressure, and chemicals).

The number of cells involved and their differing reaction speeds and stimuli are vast. They have an effect on the entire body. Any owner will know that the sense of touch is acute around the nose and muzzle. Horses use this area to communicate with other horses and now with humans. We, in turn, use this sense to help instruct the horse when it is being worked and we want to communicate direction or pace. Despite the hair that covers the horse, the skin receptors will still feel a fly landing on it and react accordingly in an attempt to remove the annoyance.

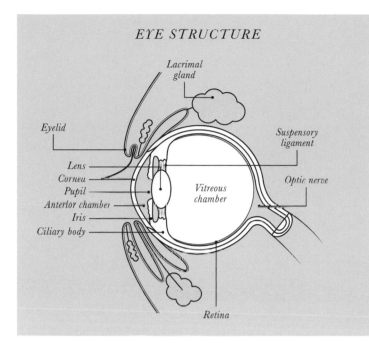

EYE STRUCTURE

Lacrimal gland

Eyelid

Suspensory ligament

Lens
Cornea
Pupil
Anterior chamber
Iris
Ciliary body

Vitreous chamber

Optic nerve

Retina

through the lens and the gel behind it, it hits the retina. The majority of the retina contains light-sensitive receptor cells, which activate responses along the nervous system, including the brain, which translates them as light/dark and associated visions. Surrounding the eye is the tough sclera, which helps to maintain pressure and protect the eye.

The eye contains a complex system of fluids, blood vessels, and nerves. Together, these support the nutrients, eye positioning, and resulting vision. The horse is protective about its eyes and frequently, for an eye examination or similar procedure, nerve blocks around the eye may be used to relax the muscles. The iris is usually colored; colors include gray, blue, gold, brown, white, and, more unusually, pink.

VISION

Horses are prey animals, not predators, and therefore have advanced eyesight. Their eyes are positioned laterally (to the side of) on their skull, thus affording a larger range of vision and greater sight of their surroundings. The eyeball itself is surrounded by a number of muscles and ligaments, which allow for the eye to move and provide extra protection to the delicate eye. The eyelids also provide protection, assisted by the so-called third eyelid. As the horse blinks, lacrimal fluid spreads across the cornea, the outer thin layer of cells on the eyeball.

When the horse views an object, light passes through the cornea, then through the anterior chamber and pupil, past the iris, and in the lens. The lens is able to change shape, thus letting the horse see closer or farther-away objects with more precision. Once the light has passed

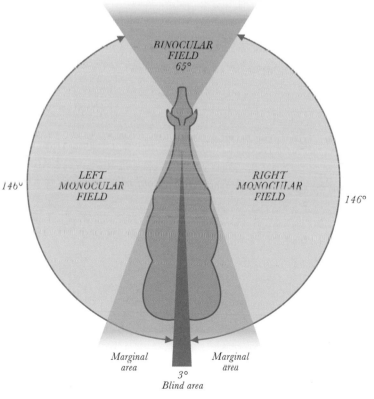

FIELD OF VISION

BINOCULAR FIELD
65°

LEFT MONOCULAR FIELD

RIGHT MONOCULAR FIELD

146°

146°

Marginal area

Marginal area

3°
Blind area

How the Hoof Developed ⟨∽

When we recall that the horse is in the order Perissodactyla (odd-toed ungulates), we can see that it has experienced significant changes in order to evolve from mammals with five toes. The foot has undergone a large amount of evolution to become the single-toed hoof that exists today.

EARLY CHANGES

The early equids walked on spread-out toes and lived in the soft, moist soil of forests and then on grasslands. The older fossils show four toes in the forelimbs and three in the hindlimbs. As the horses moved into more open lands, they needed to run faster to avoid predators, so all of the toes were lifted off of the ground except the third one, with some weight supported on the second and fourth side toes. *Parahippus* and *Merychippus* still had three toes, but the side ones were now much smaller. Over time, *Pliohippus* had one major toe and two small remains of the other toes, and the remains from *Dinohippus* show that some had one toe and others had a similar conformation to *Pliohippus*. So we can see how the modern horse hoof and limb have evolved.

Today, we still see what may be the remnants of the second and fourth toes but now refer to them as the splint bones. One theory suggests that having just one toe enabled horse weight to increase—and, indeed, the size of the horse did increase over time—and loss of muscle in the lower leg led to lighter limbs and the ability to run faster.

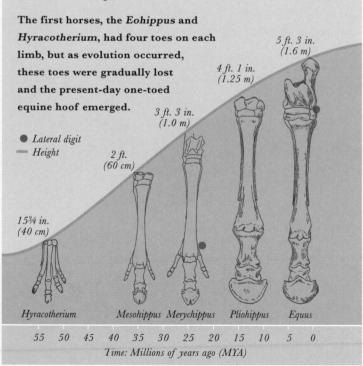

Below: *The diagram shows not only how the toes evolved but also how the limbs lengthened throughout the years as the horse grew in stature from 1 to 2 feet (30–60 cm) tall to its present-day height.*

EQUINE LIMB EVOLUTION

The first horses, the *Eohippus* and *Hyracotherium*, had four toes on each limb, but as evolution occurred, these toes were gradually lost and the present-day one-toed equine hoof emerged.

● *Lateral digit*
— *Height*

5 ft. 3 in. (1.6 m)

4 ft. 1 in. (1.25 m)

3 ft. 3 in. (1.0 m)

2 ft. (60 cm)

15¾ in. (40 cm)

Hyracotherium   *Mesohippus*   *Merychippus*   *Pliohippus*   *Equus*

55 50 45 40 35 30 25 20 15 10 5 0
Time: Millions of years ago (MYA)

PRESENT-DAY HORSES

Present-day feral and domesticated horses share a similar horse conformation, but over the years the hoof has effectively lost four digits (I, II, IV, and V) and the remaining bones, joints, and the capsule have all evolved. The result is a hoof ideally adapted for life in the harsh mountainous and plains conditions that we see the feral horse living in. This was the most appropriate foot for the wild horse's ancestors, and is a foot that enables the horse to move at speed while also supporting the heavy weight of the body.

The proximal or first phalanx (P1 or long pastern) in the limb leads down to the middle or second phalanx (P2 or short pastern), and under this the distal or third phalanx (P3, commonly called the pedal or coffin bone) resides closest to the ground. The navicular bone sits behind the coffin bone and helps to form a joint with P2. The bones assist with movement via ligaments and tendons that help to create the lever-type movements between the muscles and bones. These structures are also surrounded by the soft and harder tissues, blood vessels, and nerves that complete the hoof structure. The wall contains specialized cell and tissue types to create a tough, thick (½ inch/ 6–12mm) wall, which protects the delicate tissues within. It also provides a convenient place for attaching a horse shoe.

The modern hoof, however, is not without its difficulties. The wall can split, especially if shod poorly, and if there are injuries to the leg, some horses are prone to hoof splits—even horse management can affect whether the wall splits. Horses can also develop lameness, especially

from laminitis. The 550 to 600 pairs of primary epidermal laminae can experience inflammation, making these essential structures unable to perform their function adequately. This can result in cell and membrane damage in the areas surrounding the lamellae. Fungal infections, such as white line disease, and bacterial infections, including thrush, are also able to enter the hoof. In addition, trauma, wounds, and breakages can occur, which may in turn cause infections such as hoof abscesses.

THE HOOF CAPSULE

Long pastern (P1)

Deep flexor tendon

Plantar cushion

Navicular bone

Lateral cartilage

Short pastern (P2)

Coronary band

Hoof wall

Outer wall

Inner wall

Coffin bone (P3)

Insensitive frog

Sensitive frog

Frog apex

Soul

White line

Sensitive laminas

Above: The complex arrangement of bones, the wall, and soft tissues, such as ligaments, muscles, fat cushions, and laminae, ensure that the hoof is fit for function.

Coat & Color ❧

The domesticated horse comes in an array of colors and shades with many different types of markings.

COLORS

Black is self-explanatory and markings are white. **Seal brown** is nearly black hair with slightly lighter brown hair in some regions. **Blue roan** is used to describe a black or brown horse with a little white. The **pinto** is a horse carrying multiple colors, such as piebald, skewbald, overo, sabino, tobiano, and tovero. **Piebald** horses are black and white and their patches are irregular. **Skewbald** horses have a base color (any color except black) with white patches. The **overo** and **tobiano** have a white coat combined with any other color with a spotted pattern, while the **tovero** is a mixture of the overo and tobiano. **Sabino** also has slight spotting with high white markings on the leg.

Below: *Although horse colors and patterns may have helped the animals to blend into the background in the wild many years ago, today we often use them to identify breeds. The horse has a truly outstanding range of both colors and patterns, some more rare than others.*

EQUINE RANGE OF COLORS AND PATTERNS

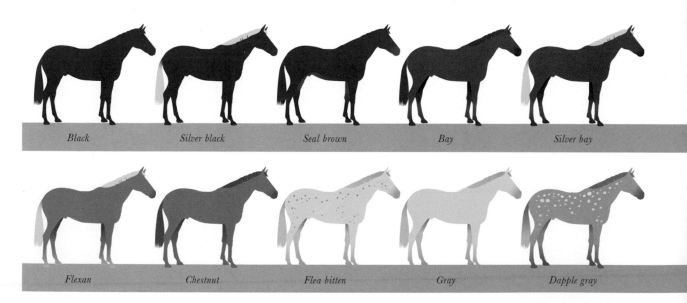

Black	Silver black	Seal brown	Bay	Silver bay
Flexan	Chestnut	Flea bitten	Gray	Dapple gray

Brown describes the darker brown to black, with brown to black points. **Chestnut** may also be called sorrel, and the tones are ginger to yellow to red. In chestnuts, the tail color is usually the same color as the coat or a paler shade. The **liver chestnut** is the darkest of the chestnuts. **Bays** contain shades of brown, ranging from red to yellow to near brown, and black points may be observed. The **claybank** horse has a darker hair color with red to yellow dun coat colorings. The golden **palomino** has flaxen or white tail and mane hair. A **buckskin** has golden hair but black points, tail, and mane. The **strawberry roan** is chestnut and white.

Duns usually have yellow hair on black skin but it can vary from horse to horse. Other markings can include dorsal stripes and leg barring, possibly a throwback to equid evolutionary patterns.

The **flea-bitten gray** is predominantly gray with flecks of darker hairs. The **gray** has black skin complemented with white and black hairs, so the gray can vary from much lighter shades to darker colorings. The **dapple gray** horse has both black and white hairs distributed in an uneven manner. In the **dapple**, darker circles are present with lighter areas in between. The **rose gray** has a more red to pink look.

Cream horses have a cream coat skin without pigmentation. Interestingly, their eyes often look pink. The **cremello** also has a pale coat but has blue or amber eyes. The **white** is one of the rare colors, having white hair and unpigmented skin with blue or brown eyes. **Appaloosa** is often used to describe spotted horses, but it is actually a breed instead of a color type. **Spotted** is the general name used for a coat color of any type with spots.

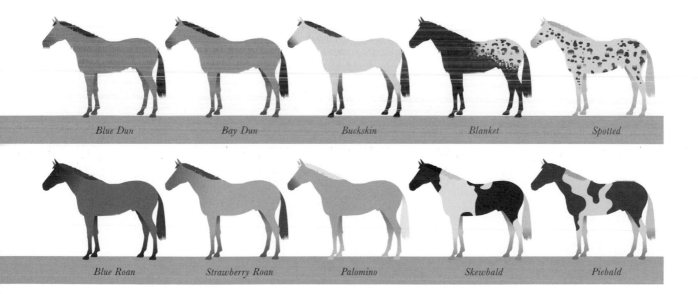

Blue Dun   *Bay Dun*   *Buckskin*   *Blanket*   *Spotted*

Blue Roan   *Strawberry Roan*   *Palomino*   *Skewbald*   *Piebald*

Faint star

Star, blaze, and snip over nostrils onto lower lip

Merged star, stripe, and snip

Star, short stripe, and snip into left nostril

Star, interrupted stripe, and snip

COLORS AND GENETICS

There are a good number of further descriptive names, and the names vary from country to country. Other people prefer to categorize based on duns, grays, base coat, dilutions, champagnes, pinto roans, and appaloosa. The true albino (white hair, pink skin, red/pink eyes) does not occur in horses, because the genetic mechanism is not present, although some people may commonly refer to white horses as albino. A genetic condition called lethal white syndrome does occur in foals, but as suggested by the name, these foals die soon after birth. Genetically, all horses are either reds (chestnut) or black. Expression of the agouti gene enables the bay color. The other coat colors are created by additional genes working off that base genetic code. Coat color is so important to some owners/breeders that some registers are available based purely on coat color instead of breed. It must be noted that due to complex genetics, offspring will not always be the same color as their parents.

Above: *Head and facial markings are useful techniques used to describe individual horses and, of course, other relations and distant cousins within Equidae.*

Development of coat color: A foal is not always born with its mature coat color. This foal, for example, will probably become a gray like its mother. The palamino foal often does not develop the diluted golden color until after it is one year old. Coat color may also depend on the season.

MARKINGS

Horses frequently have differing markings, especially on the head and legs. We describe these as white face, blaze, stripe, interrupted stripe, star, and snip on the face. On the leg, we see the stocking, sock, pastern, heel, and coronet. In the horses with a stocking, the hair is white from the coronet up to the hock. The rise of the hair then decreases in the sock, pastern, and heel until the coronet type is reached, where just a thin band of white hair is observed at the coronet.

Right: *Like many other members of Equidae, the horse can have body markings, although these may not be as extensive as the zebra. Each horse can have markings that are generalized into types of patterns.*

Below: *Leg markings are essentially based on differing degrees of color change from the foot up through the leg, ranging from none up to the stocking.*

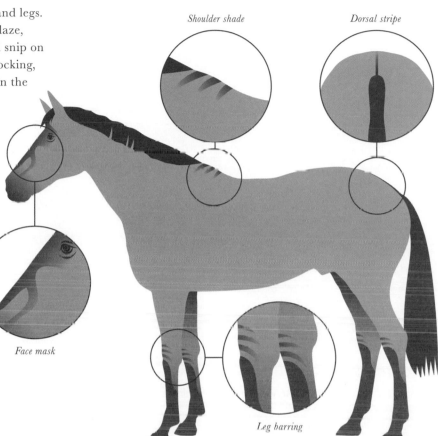

Shoulder shade

Dorsal stripe

Face mask

Leg barring

EQUINE LEG MARKINGS

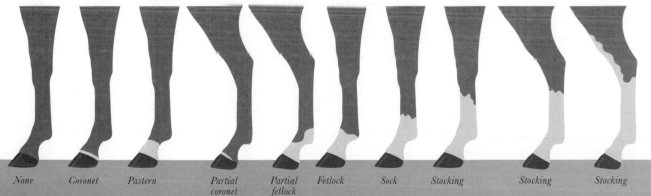

| None | Coronet | Pastern | Partial coronet | Partial fetlock | Fetlock | Sock | Stocking | Stocking | Stocking |

Equine Disease & Illness ◊

Equine health and care is an important issue and some conditions have already been discussed in this book (see the sections on reproduction, page 45, and dentistry, page 55). The key to healthcare is an observant owner/care team, a good farrier, and a good veterinarian. Equine health will also depend upon correct husbandry, such as appropriate food, clean water, a stress-free environment, well-fitting equipment, and fresh and sanitary housing. Some of the many disorders and diseases are briefly covered here.

FOALS

Caring for the foal can be difficult, because foals are more susceptible to diseases and problems. Making sure that a foal feeds is imperative. First, the foal will get vital immunoglobins from the colostrum and milk, called passive transfer, providing extra immunity to its immature and vulnerable body. Second, a failure to feed may indicate health problems in the foal or the mare. The foal can appear weak, have an elevated temperature, and may develop septicemia.

Left: *A veterinarian caring for a newborn foal. The newborn is highly susceptible to infections and diseases. While the mare's colostrum and milk will help it to develop immunity over time, foals should be checked on a regular basis for illness or even physical injury following birth.*

The foal can have a blood sample checked for immunoglobin numbers to confirm failure of passive transfer. Some owners are even able to collect and/or store colostrum from different mares to feed to such foals, or even all foals, so that they receive a good variety of immunoglobulins.

All foals should also be checked visually for abnormalities, such as birth trauma (bruising, broken bones) and congenital problems, which can range from limb problems to ocular problems, such as cataracts, a malfunctioning heart, or even more serious life-threatening conditions.

Neonatal isoerythrolysis is a particular risk for a foal. This condition occurs when the mother's antibodies start to destroy the foal's red blood cells. The foal may appear listless, have a faster breathing and heart rate, and eat less, and its urine may change color due to the blood cells being excreted. Tests using blood samples can confirm the condition, and the foal may require a blood transfusion, intensive care, intravenous fluids, and antimicrobial drugs.

A foal is generally more susceptible to digestive system problems than an older horse. Whether it is parasitic, bacterial, or viral infection, inappropriate or too much food, or so-called foal heat diarrhea, appropriate treatments must be given to make sure the horse is not at risk. These range from antibacterial drugs, electrolyte replacement, soothing ointments, and plenty of supportive care.

A foal may also be susceptible to heart problems when it is first born. In the transition from fetus to foal, the heart undergoes many changes and adaptations. Heart beat rates will differ over the first few days as these adaptations are made, and, of course, the foal must breathe for the first time.

SOME COMMON CONGENITAL CARDIAC ABNORMALITIES

Ventricular septal defect
Smaller defects usually have a good prognosis for the foal, but a large defect in the septum can result in heart failure—and *atrial septal defect*—if the foramen ovale (the septum hole required as a fetus) fails to close after birth. This means that blood can inappropriately go from the left- to the right-hand side, thereby overloading the pulmonary valve. These defects can be detected using echocardiography. It should be noted that such problems can be inherited and treatment is difficult. Close monitoring is maintained.

Tricuspid atresia
This is a serious condition with multiple symptoms, including cyanosis (turns blue), weak pulse, heart murmur, inappropriate heart conduction, enlarged left ventricle and small right ventricle, and ventricular septal defect. These foals usually do not survive long.

Tetralogy of Fallot
In this condition, four major heart problems occur at once, as also seen in humans. Due to the extreme nature of most of the heart defects, euthanasia is usually recommended.

Patent ductus arteriosus
In this condition, the ductus arteriosus shunts blood from the pulmonary artery to the aorta, resulting in a murmur. This may or may not be serious.

ADULT HORSES

Lung diseases are common in horses. A condition called recurrent airway obstruction is caused by fungal infection from hay and can damage lungs permanently. It is treated by various drugs and husbandry support, such as soaking hay to remove the dust harboring the fungus, and keeping such horses on dust-free bedding.

Colic is another serious condition in equines. There are more than 75 causes of abdominal pain or colic in the horse, and likewise a number of causes and treatments, depending on the clinical diagnosis and symptoms. These can range from fluid therapy, laxative or lubrication fluid administration, anti-inflammatory drugs, and narcotics to surgical interventions. Treatment can be expensive and after a full veterinary diagnosis and prognosis, consideration should be given to the life of the horse following surgery as well as the success rates.

Below: A horse with colic. Colic is always worrying for health professionals and, of course, owners. It is a common disorder in horses due to management problems and can cause a lot of abdominal pain and discomfort.

Horses, like most other animals, are susceptible to an extensive range of parasitic, viral, and bacterial infections. These can affect external parts of the body, such as the skin, ears, and eyes, as well as internal organs and the circulatory system. Common parasites are ascarids, strongyles, tapeworms, bots, pinworms, and habronemiasis. Appropriate pasture and housing hygiene, good grooming practice, and avoidance of overcrowding horses all play an important role in ensuring that parasites are not contracted. Treatment is usually with prescribed specialized drugs (often anthelmintics) and often accompanied by fluid therapy and supportive care.

The guttural pouch is a place where bacterial infection can often be observed, and it is susceptible to trauma (physical damage) and foreign objects. Discharge from the nostrils is a common sign of guttural pouch disorders, and it may or may not be accompanied by an observable increase in size of the guttural pouch area. Antibiotics are often given and lavage treatment may be used in more serious cases. If these do not work, surgery may be required. Fungal infections can also affect the guttural pouch, and there can be congenital disorders observed, too.

Owners should check their horses for skin problems. Naturally, horses can get wounds and scratches, but they are also susceptible to warts, ringworm, dermatitis, and even skin melanoma. In addition, they can be affected by insect bites, chemical burns, viruses, fungal infections and bacteria, abscesses, cysts, tumors, and ulcers. This is just a small list of the many diseases and disorders affecting the skin.

The Horse Genome ✑

This section looks into the world of equine genetics. Increasingly, hereditary disorders can be detected using genetic techniques and, in time, it is hoped that new therapies and cures will be developed using this information. In Chapter 1, we saw how genetics has helped inform us about the ancestry and diversity of the horse and other members of the Equidae family. Work has also been ongoing to understand equine history and ancestry using numerous genetic techniques, which will give us valuable insights into these animals. In addition, as we enter a time of genetic revolution, we are increasingly able to use genetic information for breeding purposes and to better understand health, disease, and disorders.

CHROMOSOMES AND NUCLEOTIDE BASES

A few decades ago, we discovered that the domesticated horse has 32 pairs of chromosomes (there are two of each chromosome in most cells), including two sex chromosomes. In the stallion, the sex chromosomes are X and Y, while the mare has two X chromosomes. Although the total size of the human genome is larger than the horse genome, humans have only 23 pairs of chromosomes, including two sex chromosomes (males XY and females XX).

The genome is made up of four nucleotide bases called thymine, cytosine, adenine, and guanine. Their order (sequence) makes each gene individual. Every individual has a slightly different DNA sequence overall, but it varies even more between breeds and more again between species. The horse genome is made up of around 2.7 billion nucleotide bases, depending on the exact horse, breed, and even individual. This genome is slightly smaller than the dog and smaller than the cow and the human, but it has more than 3 billion base pairs. Remarkably around half of the genome is in the same location in relation to the human; this is called synteny.

Right: *The building blocks for life in horses are DNA. These consist of four bases (A, T, G, and C, for short) held together by hydrogen bonds and a sugar and phosphate backbone. In 2006–2007, the first horse genome was completely sequenced, and it was published in 2009. It is a fantastic achievement and one that will advance knowledge, breeding, and medicine in the next few decades.*

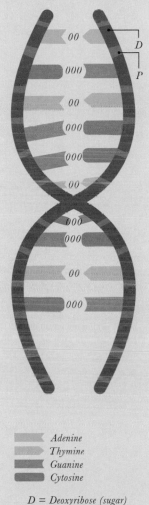

AN EXAMPLE OF HORSE DNA

D = Deoxyribose (sugar)
P = Phosphate
000 = Hydrogen bond

Adenine
Thymine
Guanine
Cytosine

SEQUENCING THE GENOME

The first horse to be completely sequenced was a Thoroughbred mare named Twilight. The effort was an international one involving more than 100 leading scientists plus their laboratory groups. In January 2007, the full sequence was revealed after years of hard work.

A quarter horse mare was sequenced in 2012, and in the same year a larger study was published showing the sequences of four domesticated horses: a Przewalski's horse, a donkey, and three fossil horses dating back 13,000 to 50,000 years. In addition, studies have looked at many different breeds, identified small sections of the genome, and have identified mutations. All this work has contributed toward the overall genetic map of the horse.

These studies, and the many carried out in the field of genetics in general, including in other animals, have paved the way for unraveling the horse genome and understanding what each gene does. Although the sequence has been identified, we still do not know what every gene or mutation does, but already many important genes and disorders have been identified.

TESTING FOR SPECIFIC GENES OR MUTATIONS

In order to test for specific genes or mutations, only a small number of cells, such as blood or saliva, are required, which makes testing especially accessible. Genetic testing is already used to help understand equine evolution and, in modern-day breeding, to understand lineage and parentage, similar to the uses in humans.

Just because a horse carries a particular gene, it does not mean it cannot or should not be bred from. It may mean that the other potential parent could be tested, because sometimes genes from both parents are needed to cause the desired or undesired effect (a recessive gene). In the case of a dominant gene, only one copy is needed from one parent. It may also help with care for the horse. In the case of malignant hyperthermia, for example, the symptoms may only show under certain conditions, such as anesthesia, so horses carrying the particular gene could be watched more carefully.

Some examples of conditions that are presently tested for include: glycogen branching enzyme deficiency (GBED), recessive hereditary equine regional dermal asthenia (HERDA), recessive hyperkalemic periodic paralysis disease (HYPP), incomplete dominant malignant hyperthermia (MH), type 1 dominant polysaccharide storage myopathy (PSSM1), and foal immunodeficiency syndrome (FIS), to name just a few. As we understand the genome further, we will also be able to test for more inherited disorders or even desired features, such as coat color.

As we look across horse genomes, we see much variation. Przewalski's horse, for example, has the highest number of chromosomes of any of the equine species. It has an extra chromosome pair compared with domestic horses. Despite the extra chromosome ($2n = 66$ in total compared with $2n = 64$ in domestic horses), it can still breed and produce viable and fertile

63   *Mules and hinnies contain 63 chromosomes*

62   *Donkeys contain 62 chromosomes*

32 *Mountain zebra has 32 chromosomes*

46 *Grevy zebra contain 46 chromosomes*

offspring when paired with domestic horses. Interestingly, we do not yet know whether the domestic horse chromosome 5 split in Przewalski's horse to become two chromosomes or whether chromosomes 23 and 24 fused to become one single chromosome in the domestic horse. We do know that the number in Przewalski's horse is always higher than other equines, so the splitting theory is favored by some. In addition, despite the differing chromosome numbers, only four alleles have been found to be specific to Przewalski's horses, again suggesting the high levels of similarity between this horse and the domestic horse.

FUTURE CHALLENGES

In addition to changes in the genetic code, we are starting to understand changes that are not induced by differing codes or mutations, known as epigenetics. Epigenetics was first formally defined in 2008, so it is a new science, but vast knowledge is already being gained. Epigenetic changes or modifications alter the way genes are expressed or not expressed. Changes lasting many generations have been observed in other species already and are being explored in the horse. In the future, it is expected that genetics and epigenetics will help not only with breeding and finding genetic disorders, but also in developing new treatments, preventative medicines, and genetic engineering possibilities.

CHROMOSOME NUMBERS

Mules and hinnies contain 63 chromosomes, which is a combination of the 62 observed in the donkey and 64 in the domesticated horse. Mules have played an important role in advancement of genetic techniques and science. Cloning has also become a real technique in the last few years. A cloned mule named Idaho Gem was created in 2003 by the Project Idaho scientists. Idaho Gem was the first cloned equine and the first cloned hybrid (being donkey and horse). Later on, two male clones named Utah Pioneer and Idaho Star were also created. Idaho Gem and Idaho Star were entered into mule races and placed well on several occasions, and were therefore fit and healthy.

The zebra species also contains differing numbers of chromosomes. Grevy's zebra has the most chromosomes at 46; plains zebras have 44 and mountain zebras have 32. Therefore it is fascinating that zebroids (zebra mated with another equine species) end up with 54 chromosomes.

44 *Plains zebra contain 44 chromosomes*

CHAPTER 3

Society & Behavior

Courtship & Mating

Reproductive behavior takes place within the harem or natal band structure, where there can be two or three breeding stallions. The stallions defend their mares instead of a territory, because the latter will change according to season as well as the availability of food, water, shelter, and other resources.

Courtship and mating in wild horses follow a predictable and typically seasonal routine, initiated by either the mare or the stallion.

MATING BEHAVIOR

Breeding stallions spend time scent-checking their mares and urine and feces markers to identify each mare's reproductive status, and they become increasingly vigilant during the breeding season. The mare will show that she is in estrus by urinating more frequently. The smell of her urine changes and acts as an olfactory signal that will trigger an innate mating behavior pattern in the breeding stallion.

The mare will usually approach the stallion to initiate sexual activity, turning her rump to him with her tail lifted and held to one side to expose her genitals, which the stallion sniffs and often shows a flehmen response to. The mare typically

INNATE MATING BEHAVIOR

Innate behavior is genetically hardwired in the brain and can be performed in response to an external or internal stimulus without the animal having any prior learning experience of that stimulus. Fixed action patterns consist of a series of predictable actions that are triggered by a specific stimulus, and all members of a species usually perform them in a similar way. The simplest form of innate behavior is a reflex action, such as a human's knee-jerk reflex when the tendon below the kneecap is tapped. The tap activates neurons that run between the knee and the spinal cord, which causes an involuntary kick action.

Instincts are more complex behavior patterns than reflexes. They are inborn, relatively inflexible, and they help the animal to adapt to its environment. A horse scratching its tail on a tree is an example of an instinctive behavior. Chemical signals, such as pheromones, perform an important function in triggering instinctive actions including alarm, mating, and nursing behavior. Instinctive behavior is often caused by the interaction of internal and external stimuli, therefore, it is also dependent on conditions in the internal environment. Courtship and mating behavior takes place when sex hormones are present in the blood supply. As a result of these hormones, the brain's hypothalamus initiates the fixed action pattern activities leading to mating. These selective responses to specific aspects of an animal's total sensory input reveal that they only respond to stimuli that their evolutionary history dictates are significant for survival and reproduction.

turns her head around toward the stallion, who will touch his nose to hers, nickering softly, and will then proceed to "groom" the mare by licking and nuzzling her along her body toward the hind legs, sometimes grasping her hock in his mouth. If the stallion has approached the mare, or following an invitation from the mare, the stallion will usually already have developed an erection and rests his head on the mare's rump as a precursory test of the mare's willingness to mate. The stallion then mounts the mare and intromission is quickly achieved.

During copulation, the stallion will usually bite the mare's neck or withers, and after some further seconds he dismounts. Both horses will then engage in close, affectionate behavior for some minutes, and they may remain closely grazing together for up to three hours.

CHOICE OF MATE

The choice of mate clearly lies with the mare, who usually permits the stallion to mount her and rejects any hopeful advances made by unwanted stallions—be they younger, inexperienced males from the same herd or outside stallions—by biting or kicking them. However, studies of some multi-male bands have identified subordinate or "consort" stallions mating the same mare, and there is considerable evidence of mares mating with stallions from outside their harem, sometimes during the same estrus period. Even mares who have been abducted into other bands will sometimes attempt to return to their original group, and mares have been known to use dense bush cover to hide from a stallion's unwanted advances.

Below left: *A breeding stallion will spend time scent-checking and grooming his mare prior to copulation. Both horses will engage in affectionate behavior for some time after mating, and they will continue grazing together for several hours.*

Below: *Once the stallion has mounted the mare, intromission is achieved relatively quickly. During copulation, the stallion will usually bite the mare's neck or withers.*

Parenting ∽

In a prey species, such as the horse, good parenting is essential in order to provide long-term nutrition and a safe, secure environment to protect the foal from predators, giving a foal the best chance to survive to adulthood and to reproduce its own offspring.

GESTATION AND BIRTH

When the time comes for giving birth, the mare will go somewhere quiet, usually during the night, to protect the newborn foal from predators. Night is also the time when oxytocin levels are higher, and one of the roles of this hormone is to induce uterine contractions. The behavioral signs that a mare is ready to give birth start when the mare appears restless, looks at her flanks and repeatedly paws, paces, and lies down, and seeks distance from the rest of the band. In the second stage of labor, the mare lies down and strong abdominal contractions commence.

During and after birth, the foal is completely silent, because to make any noise would mean attracting predators at this most vulnerable time in its life. The mare may remain lying down for several minutes after this exertion and while the placenta is expelled; when she gets to her feet, she will then spend several minutes

investigating (and occasionally eating) the placenta, and nuzzling and licking her foal, which passes chemicals from the mare's saliva onto the foal's body. At the same time, the mare ingests the amniotic fluid covering the foal's body. Now is the time when maternal imprinting takes place, the process by which the mare comes to know that this foal is hers, the one that she must defend and suckle. The process needs to take place within the first 30 minutes; after this time, the mare will reject any other attempts by the foal to suckle her or nuzzle her.

Above: *Mares generally foal late at night or early in the morning; the dark helps protect foals from potential predators and the foal should be able to stand and move by daybreak.*

REJECTION

A foal can lose its mother through death or rejection, or the mother's inability to rear it. A foal born in a free-roaming band will not survive, because it will have no immunity from infection, an essential behavioral development. A mare can reject a foal for many reasons, including fear, confusion, and invasive human intervention, and some mares have developed an aggressive reaction to every foal they have borne.

In managed breeding establishments, attempts at hand rearing a foal are time-consuming and can lead to future behavioral challenges. This is why fostering by a suitable "nanny" mare is the ideal solution.

Below: *An Icelandic purebred foal taking its first steps. There are no predators in Iceland, so foals learn to assess environmental threats.*

Foals & Early Life ❧

The horse is a precocial species, born well developed and able to move around, even at fast paces, almost immediately. It has evolved to survive by running away from predators, so a newborn foal needs to be breathing, on its feet, and able to move quickly within a short space of time.

FIRST BEHAVIORS

The foal must progress through a defined developmental sequence in order to survive and respond to the postnatal environmental challenges it faces. It gets up within 50 minutes of birth, and

Above: *An Icelandic foal and its mother. As a precocial species, the horse offspring needs to be born well developed and mobile fairly immediately in order to survive its predators.*

Left: *This skewbald pony foal and its mother will stay close for at least the first four weeks of the foal's life. In addition to providing security, the mother will teach the foal how to find food and shelter, along with important social interactions, such as how to play.*

DEVELOPMENT OF BEHAVIOR

Crucial developmental stages are not just restricted to the first few days. Species-specific expression of the horse's genotype is essential for the development of the natural behavior patterns that will ensure its best chance of survival in the wild, and the onset of these behaviors is observed in a stereotyped sequence up to the age of 11 months. If these behavioral stages cannot develop in the normal way, it can predispose the managed/domestic horse to training and other problems later in life. They can often be initiated and maintained by later events, but the greater tendency is for the response to be set up at this early developmental stage. Some of these behaviors are innate, such as sleep stages, startle, and vocalizing; others are enhanced by learning, and these include appetitive and consummative responses to stimuli encountered in the environment, avoidance and escape responses, and conditioned associations to objects and events.

Grazing is an innate behavior acquired incrementally in stages. Infrequent in the first week after birth, it begins with a short period of the foal playing with grass without ingesting it, then within the first day, it nibbles the grass and, by day three, is able to tear it with its teeth. Grazing and walking simultaneously happens only after three weeks, and full adultlike grazing and plant-selection behavior is not achieved until weeks four to six. Young foals engage in copraphagy (eating feces). It is thought to be related to the development of the nervous system, acquiring intestinal resilience, and helping the foal learn food selection.

although sight is now an active sense, the external stimuli the foal experiences cannot yet be interpreted, so it begins to bump its nose along the mare's chest, front legs, and flanks, often guided by her gentle nudging, looking for teats and essential milk that contains immune-boosting colostrum in its early stages. Suckling should commence within two hours of the foal's birth.

The body of the newborn foal takes up to 24 hours to complete all the physiological adjustments that are necessary to function in the new world outside its mother's uterus. Clinically, the full period of physical and behavioral adjustment is considered to last four days. In managed horse breeding, it is good practice to keep human interventions to the minimum during this time, limited to necessary essential medical care, and intervention practices, such as imprint training, are not advised.

FOALS AND THEIR MOTHERS

For the first four weeks of its life, the foal stays close to its mother for security and as a place of sustenance and nutrition. She is the herd member who can show it how and what to eat and the other things it needs to know in order to survive, including how to play and where to find shade and shelter. It takes its mother's lead in its behavioral responses, and also in developing its own behavioral profile from her individual experiences.

Unless the herd moves on to a new location, the foal's activity is restricted to the nursery area around its mother and the other mares with foals at foot. By the second month of life, the foal spends a lot more time with other members of the harem, or the wider herd, ready to enter the socialization phase, which lasts until the end of the third month and during which the foal develops affiliative behavior, such as playing with band mates and allogrooming.

Foals in the wild remain with their mothers until the next foal is born, when the older foal is around 12 months old, at which point she will allow the newborn foal to suckle but may reject attempts by the older foal to do the same. However, if the mare is not pregnant again, the following year she may suckle the last foal for a longer period, sometimes a number of years, although frequency typically decreases. Within two to four weeks of weaning, a foal's time budget will closely resemble that of adult horses.

Below: *As a foal develops, its mother represents a safe zone from which it can venture to explore the world a little at a time, building confidence and independence.*

This mare–foal attachment serves the youngster well; as it develops, its mother becomes a secure base from which it can go out and explore the world, gradually becoming more and more confident and independent. The foal's father and siblings, as well as other band members, will all play an important role in its social development until the foal leaves the harem group at sexual maturity.

BECOMING ADULT

Juvenile colts and fillies leave the natal band in roughly equal numbers at approximately two years of age and within a range of one to five years. Some individuals stay with their natal band into maturity, but it is important that most of the younger herd members disperse in order to avoid the inbreeding that would result from the long-term relationship between the stallions and the sexually mature mares in the harem.

Above: *In the wild, foals will stay with their mothers until the next foal is born, usually at around 12 months in age. At about two years of age, juveniles will disperse and leave the natal band to avoid inbreeding.*

Social Groups & Behavior ∽

Living in an open steppe-type landscape and as a prey species that needs to be constantly alert for the presence of predators, it makes adaptive sense to take turns at being on watch, and the many eyes of a group give a better chance of survival. Horses also sometimes need to travel long distances to seek out food, water, and shelter, so it is beneficial for the survival of the group for individual horses to have the flexibility to adopt different roles in a group, either as lookout or as the one who leads the group to better grazing. This, crucially, gives the individual a better chance of surviving and producing genetic offspring. Groups are also able to defend resources and offspring more easily, and they allow for communal care of foals and young horses.

HAREMS, BANDS, AND HERDS

Free-roaming horses live in two types of group: a harem (or natal band) or a bachelor band of juvenile colts and stallions without their own harem. One breeding group can have more than one stallion; typically the breeding stallion (two or more have also been recorded) will mate with the females in that harem; however, mares sometimes choose to mate with another stallion in the group, or a visiting bachelor—if she can escape the watchful eye of the breeding stallion— which many mares appear to achieve.

Bands are typically stable groups and small in number, ranging from 3 to 12 on average, and they occupy a home range that includes overlapping areas of other bands. Depending on the availability and location of resources and ecological conditions, bands will mix with others to a greater or lesser degree; for example, when water is scarce, several bands may

Right: *The horses found on the island of Shackleford Banks, North Carolina, formed harem groups as well as defended actual territory. This is fairly unusual and was possibly the result of a population containing a high proportion of mares.*

Below: *As a prey species the horse increases its chances of survival by living in groups or bands. The saying safety in numbers could not be more applicable as band formation allows the horses to take it in turns to be on lookout for any potential predators.*

A harem has two or three breeding stallions and a group of mares.

A bachelor band contains young colts but no mares of their own

Stallions from bachelor bands break away to form new harems by acquiring mares

join together at a water hole, or in large open areas, bands will mix together as a larger herd. In territorial equid species, such as Grevy's zebra or the kiang, stallions defend a territory, whereas in harem-forming *E. f. caballus*, the band stallions generally defend their mares.

HAREM FORMATION

Stallions obtain mares for their harem in several ways. They may obtain unguarded juveniles dispersing from their natal band or adult mares who have separated from their band or whose stallion has died or can no longer defend his mares, they may abduct some mares from a harem by raiding the group; and on rare occasions they may obtain an entire harem after defeating another stallion in a fight. Some stallions do this by first attaching themselves to a harem as a satellite male. Stallions within a bachelor band can acquire mares in any of these ways, sometimes initially by forming a bachelor alliance. In this way, bachelor groups gradually split up to form several new harems.

THE SPECIAL CASE OF THE SHACKLEFORD BANKS HORSES

Shackleford Banks is an island off North Carolina and is home to about 100 horses. Although horses as a species are categorized as nonterritorial and harem forming, these characteristics are not without exception. Some horses on the island formed harem groups as expected for their species; however, unusually, other harem stallions were also observed to defend territory as well as their mares. One of the reasons for this was a population with a high proportion of mares, which reduced the likelihood that they would need to be defended from competitor stallions; and any rivals that were present in the area were easily seen, because of the lack of vegetation. A further factor influencing the unusual territorial defense behavior was the narrow home range boundary of the island. Over time, the vegetation cover increased, and the horses on the island reverted to a nonterritorial organization.

TIME BUDGETS

A time, or activity, budget represents how an animal's day is broken down into the different behaviors on its ethogram. In feral/free-roaming horses, it is influenced by sunrise and sunset times, temperature, seasonal changes, food and water availability and quality, insect pest activity, weather conditions, and predator activity. Feeding and resting, including sleeping, account for up to 98 percent of free-ranging equids' time, although lower percentages down to 90 percent have been reported. The proportion of feeding to resting time varies according to the amount of vegetative cover in the home range; desert-dwelling horses, for example, spend less time eating. Horses graze more during the day in winter, while grazing at night increases in the summer, possibly because there is less forage in the winter, and some of their time during the day in summer is spent resting to escape from insects.

Movement makes up a relatively small amount of free-ranging time budgets, although horses are frequently moving while they eat, moving one or more of their feet with every few mouthfuls of forage. Horses will move during a 24-hour period to find preferred rest and shelter spots. There is more movement to water sources in dry conditions, and horses drink less when the distance from good forage to water is high; free-roaming horses in Australia have been reported to wait up to 48 hours before moving on to locate water sources.

Social communication and maintenance behaviors make up the remainder of an individual's and the group's time budget. Maintenance behaviors include rest, locomotion, drinking, elimination, and eating. In coat-care behaviors, the horse rolls from side to side on its back, swishes its tail to remove insects, scratches body parts that can be reached with hind hooves, and rubs its head or rump on trees or fence posts. To facilitate thermoregulation, horses will seek shade, shelter, or water and turn their hindquarters to the wind.

Interactions between group members result in a complex resource-holding matrix that can be relatively fluid, depending on an individual's motivation to control a particular resource and the overall availability of resources.

Below: *Free-roaming horses in Turkey. Surprisingly, movement over any distance forms a relatively small part of a horse's free-ranging time budget. In hot, dry conditions, such as those shown here, there will be greater movement as horses move toward water sources.*

THE HORSE'S ETHOGRAM

An ethogram is a species-specific list of behaviors that are performed by animals living in specific environments. These can be wild, free roaming or feral, and in captivity. "Normal" behaviors under these environmental conditions vary due to the availability of opportunities, which include resources that can be accessed by the animals within that environment. Human interactions regarding the availability of opportunities can impact negatively on an animal and many behaviors deemed "abnormal" are usually a product of domestication or captivity.

Ethological maintenance behaviors	Behavioral indicators include	
Safety—responding to environmental threats or perceived threats.	• The ability to orient toward potential threats, position body ready to move toward or away, often at speed.	• To be free of pain in order to do this. • The ability to respond individually and as a group by maintaining individual and herd space.
Ingestion—eating and drinking.	• To eat a forage-base diet for up to 16 hours a day. To have access to fresh water daily.	
Body care—rolling, stretching, etc.	• The ability to roll on various substrates in order to itch, remove loose hair, or coat in mud for waterproofing. • Tail swishing and kicking to displace flies.	• To have familiar herd members to mutually groom with. • To be free from pain in order to do this.
Rest and sleep—either standing or lying.	• The ability to achieve normal sleep patterns, which include dozing while standing, having locked their "staying apparatus." • To lie sternally recumbent to rest and doze. • To lie laterally recumbent, which is the only way that REM sleep can be achieved.	• Appropriate sleep can be achieved only within familiar surroundings and with familiar herd members, some of whom will stand on "sentry duty" while others lie to rest and sleep. • To be free from pain in order to do this.
Physical motion.	• The ability to freely move around their home range. Including but not limited to, walking while grazing, reacting to threat and play between herd members. • Play can only occur when members of the herd are stable and familiar. • Play patterns are innate and as such, are practiced soon after birth and continue into adulthood, particularly more so with colts, stallions, and geldings.	• Play patterns include that with objects, which reduce fear responses to novel stimuli within the environment. • Locomotor play, such as "chase and charge," and physical contact play, such as "nip and shove" and play mounting. • To be free from pain in order to do this. • Herd cohesion is strengthened by play.
Exploration—of home range and environment.	• The ability to roam their surroundings and access all available resources, including herd	members, water, favored vegetation, shade, and shelter.
Association— herd dynamics.	• The ability to establish and maintain lifelong attachments with herd members, including those that are family.	• Reproductive behavior, including courting, mating, and parenting.

Eating & Sleeping ✇

Horses who eat and sleep together in groups experience less stress than horses kept in isolation. Not only are these behaviors essential for maintaining body function and physical health, but for a prey species, such as *E. f. caballus*, eating and sleeping are social behaviors crucial for psychological well-being. In the equine hierarchy of needs, safety is the base on which all other activities depend, and horses who do not feel safe will stop eating and probably experience sleep deprivation. In their natural ethogram, horses eat, rest, and sleep in groups for safety, because there will always be a rotation of herd members to remain alert and look for predators and other dangers in the environment.

Below: *Horses tend to prefer eating young plants, so pasture and grazing management may be needed to allow pastures to regrow.*

NATURAL EATING BEHAVIOR

Free-roaming herbivorous horses graze and browse for up to 16 hours a day. As plain dwellers, horses have been seen to travel approximately 10 miles (16 km) a day, though extreme distances of up to 34 miles (55 km) have been reported, with the horses "trickle feeding" while on the move and being most active through the night and at dawn and dusk. As well as eating a variety of different high-fiber, low-energy grasses and herbs, horses ingest leaves, branches, and bark from trees and bushes, a throwback to an earlier stage of their evolution, when the forest-dwelling *Eohippus* was the size of a small dog and ate forest fare. Horses use their prehensile lips to manipulate leaves and grasses, which are then ripped away by incisor teeth, while bark and branches are gnawed. Horses will lift their heads while chewing to check for threats in the environment, and naturally living

horses eat slowly and carefully, using their senses of taste and smell to select specific herbage and reject unpleasant or unfamiliar items.

The behavior of grazing is not simply ingestive; it involves an appetitive, seeking phase and a subsequent consummatory phase that involves eating and swallowing the food.

Horses have a single stomach, the structure and function of which requires the horse to keep moving during an almost continuous intake of small amounts of food. Departing from biologically driven behavioral and nutritional requirements by restricting movement or limiting access to forage risks the onset of medical problems. Such problems include gastric ulcers, colic, and equine metabolic syndrome (laminitis is one of the symptoms of this disorder), and stereotypical behavioral disorders, such as weaving, crib-biting, wind sucking, or box walking.

Below: *Major constituents of feral horse diets are graminoids (herbaceous plants with a grasslike morphology, such as elongated culms with long, bladelike leaves; for example, grasses, sedges, and rushes), while forbs (herbaceous flowering plants that are not a graminoid) and shrubs (medium-size woody plants) play a more limited although still significant role, particularly in winter.*

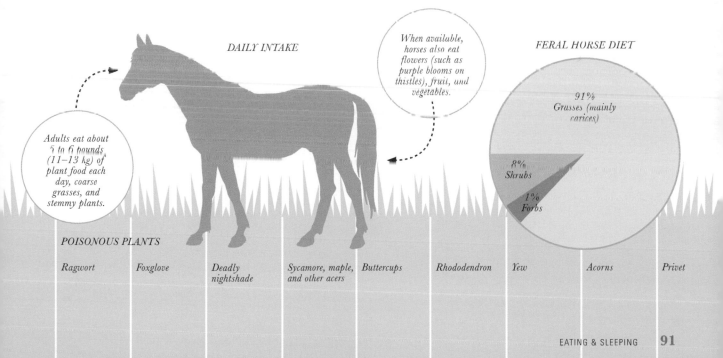

DAILY INTAKE

When available, horses also eat flowers (such as purple blooms on thistles), fruit, and vegetables.

FERAL HORSE DIET

Adults eat about 5 to 6 pounds (11–13 kg) of plant food each day, coarse grasses, and stemmy plants.

91%
Grasses (mainly carices)

8%
Shrubs

1%
Forbs

POISONOUS PLANTS

Ragwort *Foxglove* *Deadly nightshade* *Sycamore, maple, and other acers* *Buttercups* *Rhododendron* *Yew* *Acorns* *Privet*

GOING WILD IN THE STABLE YARD

In an intensively managed setting, horses should be able to eat and sleep in a daily routine that reflects the time budgets of their free-roaming counterparts. Stabled horses kept for leisure and sporting activities are often fed concentrated higher-energy foodstuffs in routines that do not replicate the natural ethogram, and isolation in single-occupancy stables adversely affects eating and sleeping behaviors that compromise mental and physical well-being.

Behavioral and welfare scientists advise that horse owners and caregivers should introduce the same environmental enrichment recognized, for example, by the conservation and farming industries and now frequently seen in many zoo enclosures. For horses, this includes feeding a variety of ad lib forage spread around the floor of the stable, scatter feeding any concentrated rations, encouraging object play as an innate motor pattern, facilitating browsing behavior by giving access to logs and branches, and enabling social interaction with herd members. Where settled groups of horses are well known to each other and there is little change in the membership, group housing in a barn system is a cost-effective and relatively low-maintenance means of achieving social interaction and physical movement.

THE BEHAVIOR OF SLEEPING

Resting and sleeping behavior in horses is divided into standing resting, or loafing, and lying down and sleeping. When grazing time increases, which in free-

roaming horses it does according to season, resting time decreases, but time spent lying down remains consistent and occupies around an hour a day. Unlike humans, adult horses are crepuscular (most active at dawn and dusk) and polyphasic; they sleep, idle, doze, or rest for 3 to 5 hours per day in short periods of time throughout a 24-hour cycle, mostly at night. Sleep disorders can have adverse physical and psychological impacts, and horses need to feel safe in order to express their normal sleep patterns, so they need the proximity of a group or pair bond.

In groups or pairs of horses who are familiar with each other, individuals take it in turns to lie down and sleep, so that there is always at least one horse who remains standing, alert for possible predators and other dangers.

Above: *Studies in the behavior and welfare of stabled horses advise that horse owners replicate the horse's natural ethogram, where possible, by providing group housing or turnout to encourage social interaction, forage feeding, and object play and giving access to logs and branches that also encourage browsing.*

Horses typically adopt different postures in different sleep stages: in stage 1, waking drowsiness, they will stand relaxed and resting one hind leg, eyes half closed and snoozing, while the stay apparatus in their front and back legs lets muscles relax without the horse falling over. In the slow-wave sleep of stages 2 to 4, they will lie down in sternal recumbency (lying on their chest and belly instead of on their side). Horses lie flat out in lateral recumbency and enter rapid eye movement (REM) sleep for about 30 minutes each day in bouts of 3 to 10 minutes, during which time it is thought that they dream, because they sometimes vocalize or kick out and make running motions. This REM phase is essential to biological and psychological well-being, and if they cannot lie flat out due to, for example, pain or isolation

anxiety, horses can quickly start to experience sleep deprivation and collapse due to losing muscle tone while standing and entering REM sleep, which is often incorrectly thought to be narcolepsy.

Horses are reluctant to lie down on certain flooring surfaces, such as concrete or rubber matting, in new environments, and in confined spaces, and traveling in horse boxes or trailers also disrupts horses' sleep. Researchers have found that when horses are transported, their REM sleep is disrupted for up to three days following travel. This can adversely affect their well-being, particularly if they have other causes of stress and anxiety. Allthough horses will normally yawn when they are tired, if they are yawning out of context—for example, repeated yawning in quick succession—it can be an indicator of stress.

Below: *The REM phase of sleep is crucial for both biological and psychological well-being in a horse. If a horse cannot lie flat out for this phase, it will rapidly start to experience sleep deprivation, which in turn can lead to stress and anxiety.*

Horse–Horse Communication ✆

As social animals living in large, cohesive groups, horses have developed sophisticated ways of communicating information to each other, about their environment, their relationships, and their emotional state. Much of the information they communicate is important for their survival, such as the location or movement of a predator, preferred grazing or watering sites, or essential escape routes. Their ability to interpret signals from other horses by sound, smell, vision, touch, or body posture is just as revealing, within their ecological niche, as language is to humans, and researchers are still learning what many of these signals mean. For example, in a study where horses had to select a bucket of food by looking at the ear direction and eye gaze of a horse in a photograph, the horses selected the correct bucket 75 percent of the time. However, when either the ears or the eyes of the horse in the photograph were covered up, selection success decreased and became more or less random. This study supports the view that horses look at other horses' eyes and ears to learn about the location of food, and this communication behavior may extend to other valuable resources.

VISUAL COMMUNICATION

Particularly minute and complex communication takes place involving the musculature of the face and neck, so that tension around ears, eyes, nostrils, mouth, chin, and neck changes according to context, visually signaling a variety of emotional and cognitive states, including fear, frustration, aggression, familiarity, curiosity, and surprise. A typical combination of signals communicating fear is a high head carriage, tense neck muscles, wide eye showing the white sclera, and flared nostrils. In resource-holding contests among familiar horses, simply the flick of an ear or the slightest turn of a head will signal to a hopeful contester that the other horse has first refusal, for example, over a patch of preferred grazing.

Horses communicate spatially by actively deviating from their current action or route in order to avoid another individual, a clear visual signal that has been interpreted by human observers as submissive behavior but may alternatively indicate individual preferences for certain group members over others, or be a choice to avoid confrontation in order to preserve energy and avoid weakening the cohesion of the group.

Right top: *Horses will greet each other by sniffing, nose to nose, regardless of whether they are familiar with one another or not. The horse's powerful sense of smell is used to detect pheromones and determine whether something is a potential threat or not.*

OLFACTION

Horses use their considerable sense of smell to detect pheromones, food, water, and other environmentally important resources or threats. Familiar and unfamiliar horses sniff each other, nose to nose, during greetings, and nose to elbow, flank, and genitals during stallion interactions. In courtship, stallions will sniff a mare's elbow and flank area, which contain high concentrations of sweat glands.

THE FLEHMEN RESPONSE

One particularly visible olfactory response is "flehmen," which horses exhibit by drawing up and back their top lip and nostrils and then inhaling to move the scent to their vomeronasal (Jacobsen's) organ, which has a high concentration of olfactory nerve cells. Flehmen is seen in foals and adult horses in response to strong or unusual smells, and stallions show a flehmen response when they smell an in-estrus mare or her urine or dung. If the stallion wants to mate with her, he will urinate or defecate on top of her excretion, and stallions also demonstrate communicative marking behavior by defecating on the dung of other males. Breeding stallions mark their home range by depositing dung piles throughout the area, often along trails used by other bands to access resources, because these are easily visible. This marking behavior is present in the majority of interactions between free-ranging male horses, and bachelor stallions may defecate in order of rank, from lowest to highest. Living in open spaces with little dense vegetation to hold scent, the complex combination of volatile substances in feces, which will remain present for several days, carries useful information about other horses' age, sex, reproductive status, and individual identity.

However, a horse may also extend its neck and curl its lips when suffering from colic, so call a vet if you are in any doubt

TACTILE COMMUNICATION

A great deal of horse–horse communication involves staying close to other horses or a preferred partner. Horses groom each other by scratching with their incisor teeth, usually in preferred areas, such as the withers at the base of the neck, the site of a major ganglion of the autonomic nervous system and found in experiments to reduce the heart rate of the recipient. Grooming maintains pair and band cohesion, an essential evolutionary adaptation in a group-living species.

AGONISTIC COMMUNICATION

Group-living horses engage in biting, kicking with front or hind feet, chasing, barging, and threatening behavior. Agonistic communication results in increased distance between individuals as an outcome of aggressive or defensive interactions. Many agonistic interactions between horses are too subtle or happen too quickly to be observed. True aggressive behaviors are avoided in favor of cooperation among the social group, because aggression uses up valuable energy, presents a risk to health because of injury, and serves to separate individuals, thus weakening crucial group cohesion.

PLAY

Play communication also involves agonistic behavior; however, an important observable difference is that ears are pinned back against the head in true aggressive encounters. Play among two or more horses involves running and mock fighting; individual horses also engage in exploratory object play, where they will investigate an object with their mouth. Two distinct interhorse play patterns are observed: "chase and charge," where two or more horses will gallop from one point to another, stop with heads up and muscular tension as though looking out

Below: *Horses can form strong bonds between one another and much of horse-to-horse communication involves tactile and familiar interaction.*

for a predator, then gallop in another direction and repeat the same pattern. "Nip and shove" involves two horses standing face to face and nipping each other's chin and jaw area, and sometimes elbows and the back of the forelegs. They may also shove into each other, trying to push the other one to the floor.

Young and juvenile horses play more than adults, and males are more playful than females, but adult horses also engage in play behaviors. Play functions to promote cognitive, motor, and social flexibility and development, and it improves muscular strength and fitness and promotes cohesion between pairs and groups. Solitary and affiliative play indicates the presence of positive affective or emotional states that humans may describe as happiness or pleasure.

Above: *Horse play is important for a number of reasons. It builds muscular strength and fitness, promotes cohesion within a group, and encourages cognitive and motor development.*

Vocalization ❧

As a prey species, horses have evolved to live in plain sight on vast, open steppe, therefore the ability to be silent to avoid drawing the attention of predators is crucial. Consequently, they are not a particularly vocal species.

DISTINCT SOUNDS

Five distinct horse-to-horse vocalizations have been identified.

Nicker: This is done with a low vocal tone and accompanied by quivering nostrils, typically between two horses who know each other, such as from a mare to her foal or between band mates. It appears to function as a greeting, often observed when one horse returns from afar or approaches a preferred partner. It can also be directed toward a familiar human, who, for example, is bringing food to the horse.

Whinny: A "Where are you?" or "Where have you been?" message, when contact with a specific companion has been lost, or when the vocalizing horse has lost sight of the others. Domesticated horses direct this auditory signal toward human caregivers, too.

Squeal: A shorter, high-pitched sound emitted during moments of social excitement or emotional arousal, typically when unfamiliar domesticated horses meet, or in free-ranging horses during resource holding contests or in courtship rituals. The squeals of dominant stallions last longer than those of subordinates and are higher pitched at the start. Both these phenomena serve to signal status and fighting ability, and they are adaptive in terms of reducing the need for the costly energy expenditure and risk of life-threatening injury that would be more probable if the stallions had to engage in more fighting behavior in order to establish resource holding status.

Left: Horses are not a particularly vocal species. Being a prey animal and often living in vast, open areas has put a stop to this. They do, however, make some sound, and five clear vocalizations have been observed and identified.

Right: Foals produce a unique sound straight from birth. It involves drawing the lips back and "snapping" the teeth together rhythmically. It is thought to be a self-soothing behavior.

Snort: Horses make this noise as a sudden, strong exhalation from their nostril, at the same time as orienting their attention toward a perceived threat, including, particularly in domesticated horses, a novel object. After the initial snort, they may typically stand and look at the object, as if deciding whether to remain or run away, and during this time, they may snort two or three more times.

Roar/scream: High in intensity and produced during extreme emotional arousal, following earlier more subtle threats that may have been ignored and appearing to threaten physical violence.

Horse caregivers have reported from their experience a greater variety of vocalizations with a range of different specific meanings.

Right: A horse will whinny when it has lost sight of others from his band or group. It is essentially sending a message to say "Where have you been?" or "Where are you?"

A recent study indicates that horses' vocalizations consist of two pitches occurring at the same time, one representing emotional arousal (calm versus excited) and one representing emotional valence (negative versus positive). In particular, whinnies were biphonated and contained two unrelated frequencies, one varying in energy and encoding arousal, and one varying in duration and encoding valence. This allows for horses to provide complex and nuanced information to other group members, alerting them to the actions or probable intentions of others and keeping a check on social interactions.

FOAL BEHAVIOR

Right from birth, foals perform a unique "snapping" behavior, typically directed toward adult horses. They pull their lips back and rhythmically clap their teeth together. It is often described as a submissive appeasement behavior. However, it is exhibited in situations where no agonistic behavior is present, with no conclusive evidence that it is appeasement, and it does not serve to stop the aggressive behavior of the other horse. It appears to occur at times of social uncertainty and is more recently thought to be a displacement, or self-soothing, behavior.

Horse Cognition

Traditionally, ideas about how horses think have rested in the realms of anthropomorphic assumptions, inferring human thoughts and emotions, or the view that all their actions and decisions take place on a mechanical stimulus–response basis in the context of their neural architecture and species-specific behaviors. However, horses in their day-to-day lives, whether free-roaming or managed, must deal with a variety of cognitive challenges, including discovering and remembering the locations of food, water, and predators at different times of year, making sense of their complex social networks, learning to perform unnatural tasks, and coping with the personality and training ability vagaries of human beings.

WHAT DO HORSES KNOW?

Horses have extremely sophisticated social cognition; they can receive, store, process, and retrieve complex information about their own herd members or field mates (though the claim that they can remember individual rank or resource holding potential has not been proven), and positive relationships with different individuals. They can do this over long time periods of months or years.

Beyond their own species, horses remember qualitative information over time about humans, places, and objects, too, and this is an important point to remember when training horses. It is advantageous to make sure they have positive learning experiences, because this learning stays with them for a long time, instead of having to deal with the behavioral fallout from a negative experience, which will last just as long and make life difficult for both horse and caregiver.

Above: *Psychological studies carried out on riding school horses showed that they displayed a pessimistic judgment bias compared to horses in more natural conditions, which displayed an optimistic judgment bias.*

Left: *Both domesticated and free-roaming horses must overcome a number of cognitive challenges on a day-to-day basis. For the managed horse, this is an extremely important point to be aware of as a human. If you are working with horses, it is highly rewarding for both horse and human that the experience is a positive one as opposed to a negative one, because this could lead to behavioral problems going forward.*

Horses also display more advanced individual cognitive abilities involving learning complex categorization and concepts. These have now been demonstrated in research experiments. Higher cognitive function confers an adaptive advantage in identifying environmental categories, such as types of food, different predators, and landscapes, and in being able to instantly classify new or unpredictable stimuli and adjust behavior accordingly.

Cognitive studies have found that when learning about the world around them, horses first learn to respond differently to diverse objects or events by a process of associative, stimulus-response learning. After this stage, they can "learn to learn," a cognitive attribute considered to be a more advanced form of learning. Using discrimination tests, the horse learns that when selecting one stimulus from a pair, it receives reinforcement (typically food), but if it selects the other stimulus, no reinforcement will be given.

HORSE PSYCHOLOGY

Human psychology recognizes a two-way relationship between emotions and cognition. In appraisal theory, individuals cognitively evaluate their environment based on characteristics, such as familiarity and predictability, which trigger an emotional experience that can be measured behaviorally and physiologically. Scientists have found that animals demonstrate emotional responses to anticipation and predictability of an event, discrepancies between expectations and reality, and the amount of control they have over their environment. Because emotional responses involve cognitive processes, this has led researchers to investigate judgment bias in a number of species. This was studied in horses, and it was found that riding school horses displayed pessimistic judgment bias, whereas horses living under more naturalistic conditions displayed an optimistic bias.

The Horse's Memory ∞

Memory is an essential element of learning. Memory systems within the brain are complex and are still being scientifically debated. Memory involves encoding, storing, and retrieving information, and a number of neural, cognitive, and emotional systems interact to perform these functions. At the level of neural communication, activation of neurotransmitter receptor cells results in long-term potentiation, which produces changes in synapses enabling acquisition and storage of new information.

CATEGORIES OF MEMORY

Two of the basic categories of memory are short term and long term. Short-term memory is easily interrupted; when this happens, the experience does not transfer to long-term memory and cannot be used at any future time to inform the animal's decisions or action. Long-term memory is more robust and can be retrieved at a later time after the information was acquired. The hippocampus, located within the brain's limbic system, is essential for the effective functioning of long-term memory, and damage to this structure makes memory consolidation impossible. Horses are more likely to store information in long-term memory when the experience is more salient, that is, when it is vividly experienced and important to them as a rewarding or aversive event, or a threat to their survival. As a prey animal, intensely fearful experiences are easily remembered, and this makes adaptive sense when survival depends on perceiving and attending to a threatening stimulus and remembering at the next encounter that it is to be avoided.

Left: *The ability to perform challenging dressage steps correctly is reliant on what is known as procedural or explicit memory. This type of memory is what enables horses to learn, in other words knowing "how" to do something.*

Memory is also classified into declarative or explicit memory ("knowing what") and procedural or implicit memory ("knowing how"). Declarative memory enables horses to retain spatial and relational information about their home range, stable yard, turnout area, objects in their environment, herd members, and significant humans; procedural memory is what lets horses learn how to organize their legs in order to canter, jump, or execute complex dressage movements successfully.

MEMORY AND LEARNING

Information previously committed to memory is best recalled in the same environment in which it was learned, a phenomenon known as place- or context-dependent memory. Memory is also state dependent, which means that the emotional or physiological state of the animal influences its ability to learn and recall information. Knowing this can help greatly when training horses. Making sure a new task or behavior is well established in one place and being sure the horse is calm emotionally will give the best chance that the new response will be effectively stored and retrieved in that context. The behavior can then be "proofed" in other places as the learning becomes established. Various aspects of domesticated living can cause stress in horses. These include inappropriate living conditions, frustration, aversive training methods, restraint, and confusing responses from their trainer. In tests, punishment has been found to result in changes that make it harder for animals to learn new tasks, even leading to psychological states such as conditioned suppression. Therefore, to achieve effective learning, it is essential to make sure that the horse's whole life experience avoids punishment, and that the horse is able to respond within its abilities to the training and management demands asked of it.

Above: *A horse's ability to learn and recall information is very much state dependent. In other words, the animal's physiological and emotional state will greatly impact on whether or not it retains a new task or behavior. The more relaxed the horse, experiencing positive emotions, the more receptive it will be to training and in turn performing, such as show jumping, racing, or at dressage events.*

The Emotional Horse ∽

Scientists have different views on whether animals, including horses, can experience or exhibit emotions. One of the difficulties is the many different words that can be used in place of "emotion": for example, feelings, mood, and mental state. Because the word "emotion" can itself be emotionally charged, the noun "affect" is often used in its place.

Below: *There have been many studies carried out on animals to determine whether they experience emotions. The results have been positive and the horse is no exception, experiencing anxiety, fear, frustration, and excitement.*

DO HORSES HAVE EMOTIONS?

How can we tell when another being, whether human or animal, is experiencing an emotion, and how can we as an external observer describe it? Humans have the capacity to report their feelings through language, but psychotherapists report that even we can have difficulty describing our own emotions, so how much more difficult it is to infer and describe emotion in a nonhuman.

The limbic system is one of the important set of interconnections in the brain where emotions are processed. It includes the hypothalamus, amygdala, and hippocampus and is referred to as the paleomammalian or emotional part of the brain that we share with other mammals. Many behavioral and physiological studies of animal emotions have led a significant body of researchers to conclude that animals are sentient and experience a wide range of emotions, such as fear, frustration, anxiety, joy, and sadness, and some of these have been studied in *E. f. caballus*.

WHAT IS AN EMOTION?

Scientists still differ on the definition of this concept. Most recently, emotions have been described as brief and transient reactions to short-term events that, over time, accumulate to form longer-lasting affective states, such as moods. Emotions are understood in different ways by human psychologists, and animal emotions can be examined in a similar variety of ways, beginning with appraisal theory, which explains how perception and interpretation of external situations in the environment result in a subjective emotional response and an action tendency to approach or avoid.

Ekman's theory of basic emotions describes emotions as adaptive responses to the fundamental function of all animals to survive and reproduce, such that when faced with the need to detect and avoid predators, or to seek and ingest nourishment, an automatic pattern of emotional response is expressed, including changes in the autonomic nervous system, behavior, and facial expression that are distinctive to a particular emotion. Animal researchers have identified a range of discrete emotions in animals, including fear, care, panic, grief, rage, lust, and play as well as joy, love, despair, happiness, and shame.

Russell's core affect construction of emotion is a useful way of looking at affect on a dimensional basis. Discrete emotions, which are not universal even in humans, are not defined; instead, feelings may be situated on two dimensions of valence (pleasure and displeasure) and arousal (active or inactive). This lets us explore emotions across a broader range without the prescriptive labels from human psychology, which may not be helpful. Scientists researching emotion suggest that we could usefully relate affective behavior to underlying neurobiological systems to explore the evolutionary basis of affective states. A recent study suggests that the operant reinforcer typically assumed to be an effect of an underlying emotion should be understood in terms of cellular and molecular events in specific circuits instead of as "vague notions of fear reduction" and are distinct from the circuits eliciting conscious feelings of fear or anxiety, thus, not "fear" but "defensive avoidance behavior."

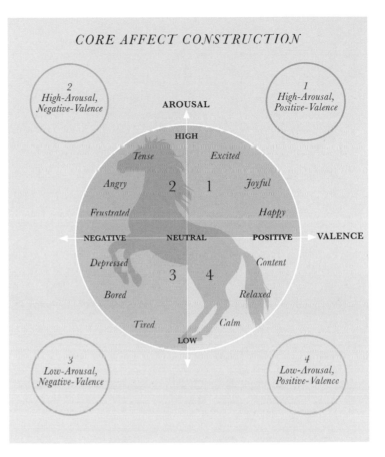

CORE AFFECT CONSTRUCTION

2
High-Arousal,
Negative-Valence

1
High-Arousal,
Positive-Valence

AROUSAL

HIGH

Tense   Excited

Angry   2   1   Joyful

Frustrated   Happy

NEGATIVE   NEUTRAL   POSITIVE   VALENCE

Depressed   Content

Bored   3   4   Relaxed

Tired   Calm

LOW

3
Low-Arousal,
Negative-Valence

4
Low-Arousal,
Positive-Valence

Above: *Russell's core affect construction lets us examine emotion or affect on a dimensional basis—feelings can be placed on two dimensions of valence (pleasure and displeasure) and arousal (active or inactive).*

How Horses Learn ❧

Understanding how horses learn, and how this differs from human learning abilities, is the foundation for effective training and a good relationship between the two species. Horses must learn about what they see around them in their environment, as well as how to perform specific actions and movements, what to approach or avoid, in order to stay healthy and to survive and reproduce. Activities performed while seeking valuable resources involve appetitive behaviors, while goal-achieving activities involve consummatory behaviors. Once a goal has been achieved, the motivation to perform the related behavior takes a long time to wane and reach the state called extinction. Even then it can reappear when the circumstances are repeated.

LEARNING TO SURVIVE

In evolutionary terms, continuous learning enables adaptation and survival. A horse that can learn to benefit from his environment and his experiences has a selective advantage by knowing what it has to do to find food, water, shelter, a herd, or a mate. Learning what to pay attention to, and what can be ignored, means that valuable energy is not wasted.

Left: *A horse will learn through operant conditioning;—in other words, that their actions have consequences. In this case, lifting one leg off the floor rewards the horse with an edible treat. This approach effectively teaches the horse what it has to do to receive a pleasant experience.*

LEARNING ABOUT THE WORLD

An essential element of learning to survive, in a world where resources may be limited by season or location, is the conservation of energy. Horses achieve this by a process called habituation, which involves learning that there is no need to respond to an event that occurs repeatedly and has no significant consequence. The adaptive function of habituation is that the horse avoids wasting energy on nonsignificant events but remains responsive to potentially significant changes in the environment that may represent a threat to safety or a resource-gathering opportunity.

Habituation, therefore, involves becoming less responsive to environment or experiences. Desensitization occurs when the horse experiences slowly increasing levels of stimulus that it learns is not worth responding to.

LEARNING WHAT TO DO

Horses learn about what is happening in their world and how to operate in their environment in order to obtain essential resources, such as food, water, playmates, and how and where to avoid potential predators. Through the process of classical (Pavlovian) or respondent conditioning, horses form associations between an unimportant or neutral stimulus and one that does have biological significance. If the neutral stimulus reliably predicts the arrival of the valuable resource, the horse will learn to respond to it. This is the way that a horse learns that the click of the barn door should always be attended to, because it always precedes and predicts the arrival of the owner or groom and the delivery of the morning feed.

Horses also learn through operant (instrumental) conditioning that their behavior has consequences. This can either be positive or negative reinforcement, where "positive" involves adding a reward such as a treat when a horse lifts its leg, or "negative" where the rider applies an aversive tool such as squeezing their leg and then releasing the pressure when the horse walks forward.

PAVLOVIAN CONDITIONING

Pavlovian conditioning is named after the Russian scientist Ivan Pavlov, who first observed and described this behavioral phenomenon in 1902. Pavlovian, classical, or respondent conditioning describes the associative learning process by which initially neutral stimuli in the environment come to predict the arrival of biologically important stimuli. If a neutral stimulus reliably predicts a stimulus that has biological significance, the animal will begin to make a similar response to this previously neutral cue that it makes involuntarily to the biologically important stimulus.

While he was researching digestive physiology in dogs, Pavlov collected their saliva, which they produced when presented with food. He noticed that the dogs began to salivate when he or the laboratory technician bringing food to the dogs entered the room, even when neither person was carrying any food. From this, Pavlov posited that some behavioral responses are automatic, reflexive, or involuntary. Pavlov tested this response by ringing a bell just before food was delivered and observed that the dogs began to salivate at the sound of the bell even when there was no food present.

Pavlovian conditioning is important from an evolutionary perspective, because it enables the animal to learn environmental cues that predict the availability of food, mates, and other important resources, and to predict potential danger, such as the threat of predators.

Wild Versus Domesticated Horses ✑

WHAT IS WILD?

The taxonomical classification of the domestic horse is *Equus ferus caballus,* a species that no longer has any wild ancestors. Groups of "wild" horses or ponies are all free-roaming offspring of horses who were once owned and managed by humans with greater or lesser degrees of intensity. Exmoor ponies in England, for example, can be traced back to a primitive but managed breed, first used as a food source, and later as pack, draft, and riding animals. Free-roaming horses in North America (mustangs) and Australia (brumbies) are the escaped progeny of domestic stock introduced by settlers, and breeds, such as the konik, have been introduced as conservation breeds to occupy the ecological niche of *E. ferus,* the true wild horse, now extinct.

A FISH OUT OF WATER?

When comparing the domesticated horse with wild species, such as Przewalski's horse, or to released/escaped groups who have reverted to wild type, it is found that social structure and behavior have changed so little that it suggests the veneer of domestication is exceedingly thin, despite the amount of selective breeding that has taken place over many centuries. The natural species-specific behaviors, for which evolution selected, remain, and domesticated horses are innately driven to express them with the same frequency and duration as their wild and feral counterparts. What has changed for the domesticated horse is the challenges presented by human management, including housing, pasturing, breeding, and training

Below: *Wild brumbies of Australia are not truly wild. They are the descendants of what was once domestic stock introduced by early settlers to the country. The true wild horse, E. ferus, is now extinct.*

practices that do not allow for expression of the complete range of adaptively evolved equine behavior.

One significant difference between feral and domesticated horses is that the majority of domesticated males are castrated and unable to mate. They cease to express innate stallion behavior, which changes the way they communicate with mares. Where mares and geldings are turned out in separate paddocks, or even individually, important opportunities to express appropriate social behaviors have been lost.

Stallions housed in isolation suffer particularly from loss of affiliative interaction and have difficulty successfully mounting or penetrating mares during mating. Without the opportunity to engage in natural copulatory behavior with a chosen male, mares are frequently mated with a stallion chosen for them, often hobbled, hooded, or twitched to ensure compliance.

STRESS AND STEREOTYPIES

Attention is increasingly being paid to the discrepancy between the horse's natural ethogram behaviors and its inability to express these in intensively managed situations. It may have no opportunity at all or there may be a mismatch between domesticated and wild time budgets, including no opportunity to exert any control or agency on its daily life, something a wild or feral horse would be doing almost constantly.

A significant challenge for domesticated horses is the regular movement between owners or livery/boarding establishments, requiring integration into a completely new

Above: *Mounted guardsmen from the Life Guards Regiment, which forms part of the British Household Cavalry. Horses did not evolve to be ridden, so without considered and gradual training, this can cause unnecessary distress to an already confused animal.*

group in which no members are familiar or familial. Individual horses find these experiences and the empty time created by a restricting environment difficult to cope with.

Horses have not evolved to be ridden, and, unless carefully and gradually introduced, the training requirements asked of them can cause states of pain, confusion, and frustration. Any or all of these factors, along with genetic predispositions, can result in stress-related behaviors, including handling and riding problems, aggression, hypervigilance, or learned helplessness.

If management criteria such as feeding the horse enough fiber for its diet are not met, we may see stereotypic behavior of various kinds that include crib biting, wind sucking, box walking, weaving, or tongue lolling. Cognitive, environmental, and social enrichment are recognized as important components in improving the domesticated horse's emotional and mental states, just as they are in zoo animals.

CHAPTER 4

Horses & People

Early Relations ∾

The first single hoofed horse was called *Pliohippus* and was the forebear of what is now recognizable as *Equus ferus caballus*, which finally made its appearance about one and a half million years before humans. *E. f. caballus* spread over the existing land bridges from the Americas and into Europe and Asia. When the ice packs retreated the land bridges disappeared, preventing any further intercontinental exchange. Around 11,000 years ago, many large mammals in North America and South America became extinct. This extinction included *E. f. caballus*, as well as other animals, such as the mastodon and saber-toothed cat. The cause of the extinction is unknown, but scientists have hypothesized it could have been due to overhunting by humans or climate change caused by the impending ice age. Horses were not seen in the Americas again until the introduction of domestic horses and donkeys by the Europeans.

The horse had evolved into three distinct and primitive types: the Asiatic wild horse; a lighter type called the tarpan, which predominated in Eastern Europe and the Ukrainian steppes; and the slower, heavier horse of northern Europe, from which current European breeds of heavy horse derive.

EVIDENCE OF DOMESTICATION

To discover when horses truly began the process of domestication, researchers have studied bone structures found in archeological sites. Bones of a smaller size and an increase in variability are considered to be signs of domestication. Variability reflects breeding under man-made conditions, and a smaller size is thought to be due to penning and diet restriction. Researchers also looked for evidence in burial sites, such as tools that may have been used for tack, and in art and adornments of items used by humans.

Below: *This is one of the six surviving horses out of the original seven that drew Surya's chariot at the thirteenth-century Konark Sun temple, dedicated to the Hindu sun god Surya, in Puri, in the Indian state of Odisha. All the horses were built from stone, as was the chariot, and on a grand scale—the 24 chariot wheels alone are 12 feet (3.7 m) in diameter.*

There is plentiful evidence for horse domestication in the Eurasian steppes. Horses appear more frequently in Eurasian Paleolithic cave art than any other animal. There have been horse head maces found that have dated to as early as 4200 BCE. This suggests that humans and horses had a special bond, although complete domestication may not have started just yet.

BEFORE DOMESTICATION

Prior to domestication at the end of the Neolithic period, the main relationship between humans and horses was that of hunter and prey. The favored hunting strategy during the ice age, as cave drawings in France and Spain attest to, was thought to be the driving of entire herds over cliffs, which was far easier than the isolation and pursuit of one animal. However, it has more recently been suggested that the horses were on seasonal migrations.

Studies of tooth cementum show most animals were killed during the warmer months of the year; perhaps they were driven along the bottom of the ridge, then held in the naturally formed corral before being killed with spears. Further investigation also reveals that few of the bones have cut marks, so they probably had not been fully butchered. Perhaps the hunters could salvage only some of the meat and left the rest to scavengers, or it has been hypothesized that horse skin was prized rather than meat.

Above: *The Terracotta Army in Xi'an, China. This is the mausoleum of Qin Shi Huang, the First Emperor of China, who was buried in 210–209 BCE along with a collection of terracotta sculptures representing the soldiers and horses in his many armies. Their purpose was to protect the emperor in his afterlife.*

The First Herders ⌘

Domestication in itself raises great debate among experts as to the exact and definitive point in time at which the horse became a domesticated animal. Was it when the horse provided a food source for nomadic tribespeople or does that turning point of the human relationship with the equine begin when the horse was ridden and used as a pack animal? At what point does exploitation become domestication? Should domestication involve actual evidence of selective breeding and offspring being born in captivity? Evidence for domestication is drawn from a number of sources, including the archeological remains of the teeth and skeletons of ancient equids. Further evidence is gathered from changes in the geographical location of equid remains, where there are discoveries of artifacts and bones in areas that did not previously have an equid population, as well as other artifactual evidence that indicates a change in the relationship between human and equid; for example, the internment of horses in graves as sacrificial offerings with their warrior riders.

Below: *Detail from the Standard of Ur, a small hollow box dating back to 2600 BCE. It was discovered in the 1920s in one of the largest royal tombs in the ancient city of Ur, Iraq. Mosaic scenes depicting man and equid together have been inlaid with lapis lazuli, shell, and red limestone.*

EARLY DOMESTICATION

Over time, early nomads, such as the Aryan tribes who occupied the Eurasian steppes bordering the Caspian and the Black Sea, began the process of domestication by herding horses, with groups being kept closer to settlements. This made the selection and procurement of food easier than hunting wild groups. This took place around 6,000 years ago and it was the terrain that partly dictated the horse as the animal of choice, because harsh conditions meant other species were less well equipped to survive on the limited available herbage.

A herd of horses could, therefore, provide a nomadic tribe with food from their meat, and the hide was used to make clothing and tents. Mares gave milk, which was often fermented and gave rise to *kumis*, the fiery beverage of the Central Asian steppes that is still drunk to this day. A mare's milk has a relatively high sugar level compared with the milk from other species, hence the resultant alcoholic content via fermentation. Herodotus described the Scythians processing mare's milk in his fifth-century *Histories*.

Evidence from the area of the Botai people suggests they were possibly the first to gradually domesticate horses for various uses, including milk, meat, and transportation. In Krasnyi Yar, Kazakhstan, archeologists using magnetic sensors discovered a set of post molds in a circular pattern, suggesting fences for holding animals. Soil tests revealed high levels of phosphates and nitrogen similar to levels found in manure. Horses were the most likely livestock at that site, so it was hypothesized that this was horse dung and dated to about 3500 BCE, suggesting that horses were kept corraled there, possibly for milk and meat.

There is archeological evidence to suggest that horses were domesticated by the Khvalynsk culture of the Eurasian steppes as early as 4800–4400 BCE. Certainly, it is clear that horses had some kind of symbolic value, because they were included in funerary rituals along with

Right: *The royal tomb, known as Arzhan-2 in the Republic of Tuva, Russia, dates from the seventh century BCE. The unknown Scythian monarch was buried with 14 horses. A symbol of wealth, the Scythians revered their horses and they were often buried in or near human graves.*

sheep and cattle, which were herded by the people of Khvalynsk. The excavation of a cemetery area in the Volga region of Russia uncovered 158 prehistoric graves, and the remains of animal sacrifices were discovered in 22 of these. In some of the graves, sheep and cattle appear to have been ritually slaughtered, and their heads and hooves interred. This form of sacrifice is known to have been used in later funeral ceremonies on the steppe; the hide would be buried along with the head and hooves, with the animal's flesh being used as some kind of feast. The Khvalynsk graves are the earliest known occurrence of this custom, and they also give us an interesting indication of the horse's role in this early society, because horse leg bones were found in ten of the graves. Because no bones of wild animals were found, we can infer that horses were considered, along with sheep and cattle, as domestic animals worthy of sacrifice. Horse bones have also been found in the graves of the contemporaneous Samara culture, along with horse carvings above the graves, which is another indication of their value.

It was not, therefore, such a great leap for tribespeople to begin to ride the more tractable animals, which made control and movement of the herds within this nomadic lifestyle easier. This also enabled tribes to range farther afield then they had been able to do previously.

Right: DNA sampling from horses in eight countries in Asia and Europe has shown there to be a genetic diversity that suggests horse domestication spread from the Eurasian steppes partly through the introduction of wild horses within herds.

DNA EVIDENCE

Research in 2012 by a team from Cambridge University, England, combined two opposing schools of thought on the domestication of horses. Archeological evidence is conclusive on the fact that horses were tamed in the western part of the Eurasian Steppes: southwest Russia, Ukraine, and west Kazakhstan. However, there is additional evidence to suggest that domestication was happening in many other places across Europe and Asia, too. This suggestion was based on evidence taken from mitochondrial DNA samples. Sampling from 300 horses from eight countries in Asia and Europe shows a genetic diversity that indicates that domestication spread from the steppes, partly through the continued use of wild horses within established herds, and this is perhaps where some of the confusion has previously lain.

Herders continued to introduce wild horses, mainly mares, into their established groups, and the DNA sampling demonstrates that horses were being domesticated in many other locations as well as Eurasia. Thus began the move away from a relationship with the horse of simple nomadic herder and hunter to one of captor and domesticator, where the horse remained in captivity effectively and bred within those confines.

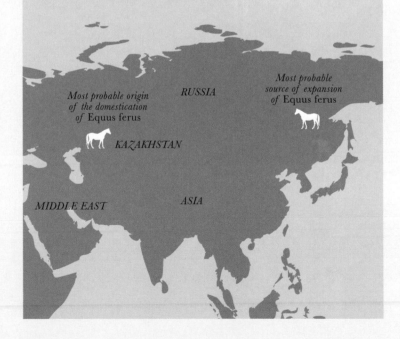

THE PROCESS OF DOMESTICATION

Once humans began to understand the horse as a pack animal and a riding animal and not just a food source, the whole relationship began to change. Initially, more docile horses were selected as suitable mounts for the control and movement of the herd. It must have been an interesting moment when these groups first realized they had discovered a conveyance that not only would allow for them to control their stock but also to hunt other species—and that could carry items heavier than they could carry.

Thus began the process of domestication in which taming and training the horse relied on sufficient skill and knowledge gained from observation of equine behavior and experience of

handling these flight animals. The horse became a beast of burden, a hunter, a war-horse, and a champion of agriculture, a role it maintained for centuries, all the way to the mid-twentieth century. It also remained and still is to this day, a food source in certain countries.

It has been suggested that of all the animals that have been domesticated, the horse is the most significant to our history. Anne McCaffrey, the famous American author, has described the horse as "the noblest, bravest, proudest, most courageous, and certainly the most perverse and infuriating animal that humans ever domesticated." This early transformation from prey animal to domestic partner shaped some of the physical and behavioral characteristics of today's modern horses, as scientists have discovered.

Above: *Once humans realized the horse was not just something to hunt and herd, a new relationship between human and horse developed. The domesticated or managed horse, with the correct training, could be used to work the land, for transport, as a beast of burden, and as a war-horse.*

Many Uses: Early Horses ∽

Early uses of domesticated horses were many and varied. Members of the Botai culture were foragers who adopted horse riding in order to hunt the abundant wild horses of northern Kazakhstan between 3500 and 3000 BCE. The advantage of horse riding made it much easier to manage herds and hunt wild horses. The Botai at this time had no wheeled carts, so it is thought they learned to ride horses first. Botai sites have no cattle or sheep bones; the only domesticated animals they had were horses and dogs, and fragments of animal bones that have been found in Botai archeological sites comprised 65 to 99 percent horse bones. The Botai developed a form of hunting where entire herds of horses were slaughtered by hunters in hunting drives, and this ability to hunt horses in large numbers led to more permanent human settlements.

Below: *A young boy on horseback in Saty, Kazakhstan. In modern-day Kazakhstan, the horse is still highly valued. Horses are most commonly used for riding and horse meat.*

INCREASING USE

As the process of domesticating the horse progressed, humans quickly learned the value of the horse as a mount, a beast of burden, and a draft animal. As a mount, the horse enabled humans to continue to lead a nomadic lifestyle and herd horses that could still be employed to provide a source of food. Humans also began to understand that horses could be harnessed to rudimentary vehicles and used to transport items, too. They also had the ability to carry messages and travel to other areas of the world and interact with cultures that had previously been unreachable.

Humans went from being able to travel at 4 miles per hour (6 km/h) to being able to travel at up to 35 miles per hour (56 km/h). Horse-centered cultures slowly began to emerge. As both a ridden and driven animal, the horse had a value in combative situations in addition to providing the ability to hunt other animals for food. In places where agriculture was important, horses were used as tools to help with work. In places where travel was important, horses were used to pull chariots or for riding.

The horse's trainability and ability to carry someone on his or her back quickly became an important and needed quality, and in many cultures, riding a horse became associated with power, royalty, and strength. The invention of riding in Eurasia would eventually lead to the discovery that the entire continent could engage as an interacting single world, and in this way horses facilitated early steps toward globalization by assisting different cultures to interconnect.

Above: *Assyrian relief sculpture at Nimrud— ca. 865 BCE— illustrating the importance of the horse in facilitating more productive hunting techniques. Here, we see King Ashurnasirpal II hunting lions from a horse-drawn chariot.*

EVIDENCE OF USE FROM ARTIFACTS

There is great debate over whether the horse was used first as a driving or a ridden animal and archeological evidence is inconclusive on this point. Some horses were ridden using crude and rudimentary bits, and evidence of this can be found from wear on the teeth of skeleton remains. However, not all horses were ridden using bits, so marks on the teeth are not conclusive one way or the other. Horses may have been put to the plow with bits used to control them in the field in hand instead of by being ridden, so wear marks on the teeth again are not necessarily conclusive of a ridden role; it could indicate an agricultural purpose. Some theorists believe that these ancient horses were probably too small to be ridden.

Much of the horse's cataloged history of domestication is deduced from what may only be called hard evidence, which is the survival of artifacts that directly point to the horse's role, and necessarily some of the best recorded finds are linked to the use of the horse in warfare, suggesting, perhaps erroneously, that activities requiring the use of harness were the first domesticated use.

One team of scientists, including the eminent Professor Leo Jeffcott, a clinical orthopedic researcher and former Dean of the Veterinary School at the University of Cambridge, England, have been trying to reach a defined point on this by looking at other anatomical changes within the horse's body, and this involves looking at the bones of the spine. It has been established that ridden horses undergo changes to the thoracic vertebrae on top of which the rider sits, and these include new deposits of bone and overriding dorsal spinal processes. These changes are observed in the modern riding horse and can also be found in the skeletal remains of ancient horses. But at this juncture, experts are yet to agree on whether the horse was first ridden or driven.

Above: *Before the invention of tractors and cars, after the ox, the horse was the farmer's most valuable asset and an intrinsic part of everyday worklife. The horse rapidly became an essential part of rural economy, enabling the farmer to work the land with greater productivity.*

SKILLS NEEDED FOR DOMESTICATION

Domesticating such a flighty animal for any role requires the use of experience and skill, and it is thought that younger horses were initially used, because continued exposure to people allows for a horse to become used to many more experiences and objects. Additionally, many ancient peoples already had a history of taming wild horses that is distinct from domestication of the species, but nevertheless provided useful grounding when humans decided they wanted to harness the horse's abilities.

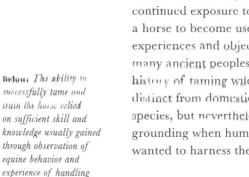

Below: The ability to successfully tame and train the horse relied on sufficient skill and knowledge usually gained through observation of equine behavior and experience of handling these flight animals.

THE MANY USES OF THE HORSE

With domestication, the role of the horse as a possible food source did not go unused. Horses were still killed for their meat and hide, although perhaps not in the same numbers as they would have been when they were the sole food source, now that it was possible to hunt other species from the back of the horse. It is probable that when breeding began in captivity, the younger horses were handled from birth and put to work while older, redundant adults were killed for their meat and mares used to provide milk. Agricultural work was generally limited, because it could be carried out by oxen, which was probably a more suitable choice of workmate at this stage in time, especially as the ability to control a horse in terms of equipment and knowledge was still at a rudimentary stage.

Lucinda Green, the former international event rider, has often said that no species does as many things for humankind as the horse, and the key early uses of the horse following domestication are remarkably still in evidence today across all the different sectors of the world. The biggest difference now, of course, is the use of the horse for sport in the Western world post industrialization, and the redundancy of the horse from its former role in warfare, in which the horse was pivotal in shaping the world for more than a thousand years. The aptly named John Trotwood Moore, the American historian and journalist, famously once said, "Wherever man has left his footprints in the long ascent from barbarism to civilization, we find the hoof print of a horse beside it."

Riding Horses: The Story of Tack ∽

With the domestication of the horse came the essential development of tack and equipment, beginning with a simple cloth over the horse's back to act as a saddle; the fixed wooden tree designs that could distribute a rider's weight more evenly came much later on in the two centuries BCE. Evidence of early saddles can be seen in the artwork of the nomadic tribes of the steppes and the Assyrians and ancient Armenians dating back to the time of King Ashurnasirpal II (reigned 883–859 BCE). Stirrups were not a feature of these early saddles. They appeared much later on, at a similar time to the treed saddle, which would make sense as up until that point, there was no real secure attachment point for them. Comfort while riding was not a point lost during early riding attempts; there is evidence that some saddle cloths were padded or cushioned to provide something pleasant to sit on.

Tack and harness developed from two key drivers: the requirement to control the horse and use it for a particular purpose (be it ridden or driven), and the dictates of available materials. So although there is some commonality among early equipment, there are regional differences.

Left: *This detail of a gypsum wall relief from the North-West Palace at Nimrud (Kalhu) in modern-day Iraq, dating back to 865–860 BCE during Ashurnasirpall II's reign, clearly shows the depiction of reins and a harness on the foreground horse.*

BRIDLES

Horses can be controlled by the nose, so it is probable that the earliest bridles were halters and the use of bits came along later. These bridles or halters were made of rope, rawhide, or sinew. Early bits were made of horn, wood, rope, or hardwood. The development of metal bits did not occur until around 1300 to 1200 BCE. Disk-shaped antler "cheekpieces" appear to have been an ancient predecessor to a modern bit shank or bit ring. The earliest recorded evidence of the use of metal bits falls within the Bronze and Iron ages, supported by archeological remains discovered in Luristan, now modern Iran but then ancient Mesopotamia.

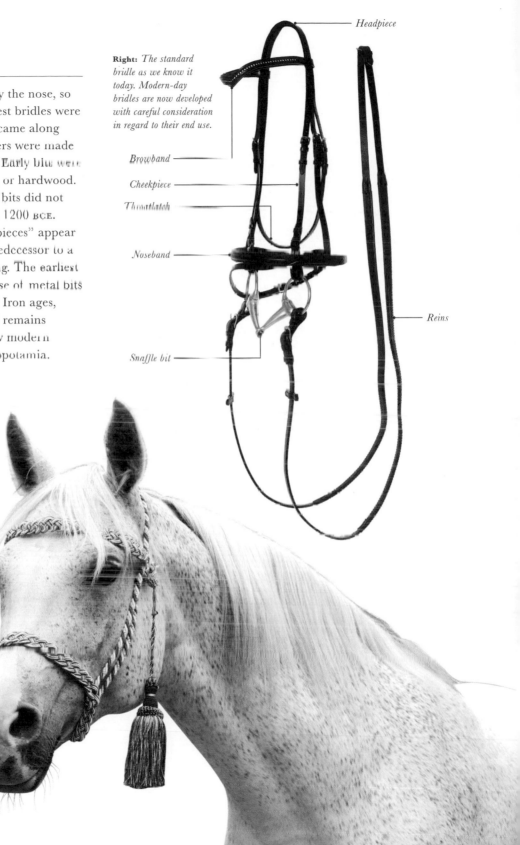

Right: The standard bridle as we know it today. Modern-day bridles are now developed with careful consideration in regard to their end use.

Headpiece

Browband

Cheekpiece

Throatlatch

Noseband

Reins

Snaffle bit

Right: Early bridles were, in fact, halters, usually made from rope, sinew, or rawhide. They relied on the ability to control the horse via a noseband as opposed to the mouth area with a bit. This is a modern day example of an Arab sporting a rope halter.

Above: *Once humans learned how to stretch leather over a wooden frame, it paved the way for the development of the early treed saddle.*

SADDLES

By contrast, saddles took a lot longer to develop, beginning life as soft pads or cloths on the horse's back and secured with a piece of rope or nothing at all, reliant on staying in place with the weight of the rider. It was the development of leather dating as far back as 1300 BCE that was the real game changer, and at some point, humans learned how to stretch leather over a wooden frame and thus began the origins of the early treed saddle. The use of leather was widespread across different continents, after all, humans had been using hides and skins for many years for clothing and to create shelter. Pieces of leather have been discovered in Egypt that date from 1300 BCE but it is known that peoples in Asia, North America, and Europe were all developing skins into leather. The Greeks were also using leather in 1200 BCE.

A treed saddle, albeit rudimentary, gave the rider greater stability and security and was also better for the horse's back, because the wooden tree ostensibly supports the weight of the rider from resting on the spinal processes. It was at this juncture that there was a point of attachment for stirrups. The first reported use of a stirrup or its equivalent is in India around the second century BCE. This was not so much of a stirrup as a simple rope loop for the big toe; Indian riders rode barefoot due to the climate. Stirrups are reputed to have been developed by the Chinese during the Jin dynasty about the first century, but their need was identified much earlier by a Byzantine emperor who mentioned the requirement for such a piece of equipment in a military manual dated 580 CE. The stirrup was initially developed in isolation as a sole stirrup for getting onto the horse by a nomadic group known as the Sarmatians. The record of the first pair of stirrups is in a Jin dynasty tomb near Nanjing dating back to around 322 CE, and such was the success of this development that its popularity spread throughout the steppes of Asia and farther afield. The security that two stirrups gave the rider led to the origins of horsemanship and gave these early peoples greater options with their horses than they had previously enjoyed.

Ironically, in recent times, in certain equestrian circles, there has been a move away from the use of treed saddles and back to saddles without trees. The idea is to give the rider a more natural experience; however scientific research suggests these modern treeless saddles actually cause back problems in horses.

DEVELOPMENT OF TACK AND HARNESS

Historically, the development of tack was a natural consequence of domestication; however, rudimentary tack is not a thing of the past, far from it. In developing countries, where the horse and other equids are still employed to work in economically poor communities, tack is still made from whatever materials are available. International charities, such as The Brooke and World Horse Welfare, spend a lot of time in the field educating local people in how to have a healthy working animal. Ill-fitting tack made from the wrong materials causes pain and discomfort to the horse and can result in wounds and sores, which mean that the working life of that animal and thereby its owner is reduced, maybe even curtailed. Both these charities provide immediate aid to horses that are suffering and transport saddlers to these countries as part of a continuous program of education. Primitive tack is not, therefore, just a museum curiosity from a bygone age.

The development of saddlery throughout the centuries has always been driven by the key principles of locality, which influenced the choice of materials, design, and the traditions of the local people. The job for which the horse was employed was also a crucial consideration; for example, a Western saddle was designed so that a cowboy can ride for days at a time. Compare that to the style and cut of a modern sports saddle specific to a discipline, designed to be used for only a few hours at a time. Each country developed its own variation on the same theme based around two generic saddle types; the Hungarian, on which the current modern English saddle is based, and the Moorish, on which the Western saddle is founded.

Below: *On the left is a standard English dressage saddle and on the right is a Western trail saddle. They are both designed for different riding disciplines and saddle time — for example, the Western trail saddle is commonly ridden over a long period of time, so it has to be comfortable.*

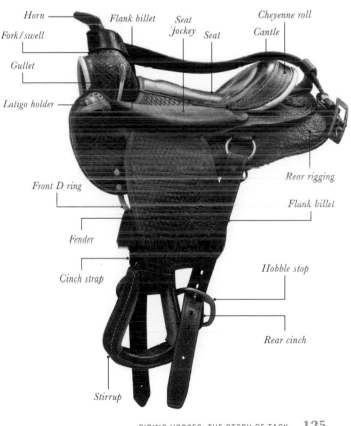

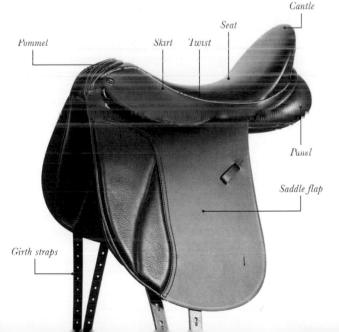

The History of Training ∞

FIRST WRITINGS

The forefather of all horse training is considered to be Xenophon, an Athenian historian and soldier who lived around 430–354 BCE. Based on riding in parades and cavalry movements required during mounted warfare, the earliest known work on training is Xenophon's treatise on the ridden training of horses. First published in Florence in the mid-sixteenth century as *The Art of Horsemanship*, it sets an esthetic tone that has continued to the present day. Historical records tell us that Xenophon spent almost his whole life with horses, and most modern commentators believe that his teaching model still holds true today. In his writing, Xenophon acknowledges an otherwise unknown horseman, Simon of Athens: "To quote a dictum of Simon, what a horse does under compulsion he does blindly, and his performance is no more beautiful than would be that of a ballet dancer taught by whip and goad." Xenophon eschewed the use of violence in horse training, recognizing that you "must refrain from pulling at his mouth with the bit as well as from spurring and whipping him. Most people . . . by spurring and striking, scare them into disorder and danger."

We could perhaps hazard that the forefather of horse behavior is Alexander the Great, educated by the philosopher Aristotle and heir to the Greek kingdom of Macedonia. As a child, he saw his father Philip and a group of men struggling to control a horse, Bucephalus, who was rearing and unmanageable. Alexander realized that due to the angle of the sun, the horse could see his own shadow moving and was afraid of it. Alexander took Bucephalus, turned him toward the sun, and he became calm and manageable.

Below: *A statue of the great Xenophon of Athens in front of the Austrian Parliament building on Ringstrasse in Vienna, Austria.*

MASTERS OF THE CLASSICAL TRADITION

The development of the use of small firearms in the fifteenth century required horses that were more maneuverable than their predecessors, whose job had been to gallop in straight lines and perform simple movements. Xenophon's works were rediscovered and the interest in classical training was reborn when Frederico Grisone founded the first modern riding academy in Naples, Italy, in 1532, where the equestrian skills and manners learned by young noblemen were considered to be essential social graces. Grisone was known for using brutal training methods on his horses, and one of his pupils, Giovanni Pignatelli, is believed to have developed the curb bit, "invented to be used as a tool of force."

Grisone's pupils set up riding academies throughout Europe, including in the seventeenth century his pupil Antoine de Pluvinel de la Baume, who opened an academy in Paris and published *Le Manège Royal*, written as a dialogue between himself and his pupil, the future Louis XIII. De Pluvinel is significant in the history of horse training for departing from Grisone's harsh methods, aiming instead to achieve compliance through kindness and rewards. In England, William Cavendish, later the Duke of Newcastle, wrote *A General System of Horsemanship in all its Branches*, and is credited with inventing the draw rein.

The most famous of the post-Renaissance masters followed de Pluvinel's humane school and taught in France in the early eighteenth century. François Robichon de la Guérinière formulated a clear and precise system embodying all the classical principles, which formed the basis of training at the Spanish Riding School of Vienna well into the twentieth century. He published *Ecole de Cavalerie* in 1729, continuing Pluvinel's work and shunning Grisone's brutality and Newcastle's gadgets.

Below: *In 1532, Frederico Grisone founded the first modern riding academy in Naples, Italy. Equestrian skills and manners were considered essential social graces for young noblemen, so Grisone's academy thrived.*

Below: *The many elements that constitute the curb bit. The bit sits in the horse's mouth and is attached to the reins. It helps the rider to communicate with the horse and allows for a degree of control.*

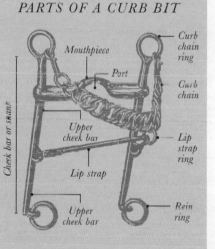

PARTS OF A CURB BIT

Mouthpiece
Port
Curb chain ring
Curb chain
Cheek bar or shank
Upper cheek bar
Lip strap
Lip strap ring
Upper cheek bar
Rein ring

By the time of the fall of Napoleon in 1815, there was nobody left who could pass on the skills taught by de la Guérinière. The elite core of instructors of the Cadre Noir was established at Saumur shortly afterward and continues to the present day. Two notable French individuals, James Fillis and François Baucher, also contributed to the development of horse training in Europe in the nineteenth century. Neither of them was well off and both had to demonstrate their abilities in the circus ring instead of in the elite academies. Fillis modified some of Baucher's training errors and was renowned for being able to ride any horse impressively with minimal effort. Wider recognition came when, at the age of 63, Fillis was invited to join the Imperial Russian Cavalry School as Chief Ecuyer, where he was lauded for training 350 remount horses every year for ten years without injury to any horse or rider.

In Spain, despite its imperial wealth and a stock of the best riding horses in the world at the end of the Renaissance, it was Austria and its capital Vienna that had the most influence on the development of training, connected by the Spanish Hapsburg dynasty and its founder Charles I. The Spanish Riding School of Vienna is so called, not because of any Spanish style of riding, but because the famous Lipizzaner horse used by the school owes much of its foundation blood to the Andalusian breed, reinforced by selective introduction of other European breeds and Arabs. The influence of the Viennese school also spread to the German states in the nineteenth century and resulted in Gustav Steinbrecht's book *The Gymnasium of the Horse,* classed alongside de la Guérinière's as one of the greatest works on training horses.

Connecting us from the late nineteenth century to the formation of the Fédération Équestre Internationale (FEI) in the early twentieth century, General Albert Eugene Edouard Decarpentry was an enthusiastic pupil of François Baucher. He served as a cavalry commander in World War I and was second in command at the Cadre Noir from 1925 to 1931. In 1949 he published the book *Academic Equitation,* considered by dressage experts to be the most important contribution to classical training in the twentieth century. Decarpentry went on to become an international dressage judge and president of the FEI dressage committee.

MODERN HORSE TRAINING

The latter half of the twentieth century saw the training of horses, or dressage (from the French *dresser*, "to train"), becoming a competitive sport in its own right, governed at the highest level by the FEI, who describe the object of dressage as the development of the horse into a happy athlete, demonstrated by free, regular paces, lightness of movements, engaged hindquarters, a lively impulsion, and an unresisting acceptance of the bit. Progression in performance is set out in a training scale that originated in a German cavalry training manual; its elements are categorized as rhythm, suppleness, contact, impulsion, straightness, and collection. However, there is no consistent standard by which to measure these principles and professionals and enthusiasts alike are calling for an alignment of training and scoring methods. Competitive dressage has taken on a form and style of its own, and present-day discourses reveal a tension between classical and modern training methods, reflected in the emergence of equitation science where the study of human–horse relationships have identified an emerging cultural paradox between the traditional and the contemporary.

NATURAL HORSEMANSHIP

In the 1980s, a collection of training techniques referred to as natural horsemanship became popular, ostensibly as a departure from traditional techniques criticized as unnecessarily harsh or forceful. Natural horsemanship claims to train horses based on their natural instincts and using interspecies communication. It uses techniques based on the application of pressure, which is released when the horse performs the desired behavior.

Natural horsemanship is criticized by traditionalists for merely rebranding older techniques, such as lunging, and by practitioners of "positive" training methods (see below) for using techniques that have the potential to hurt or frighten the horse. In scientific terms, the effects of this method are based on the application of negative reinforcement in just the same way that traditional methods utilize the application and release of pressure from a leg, a whip, or a bit.

CLICKER TRAINING

At the same time that natural horsemanship was becoming popular, another training method, based on the psychological theories of operant and respondent conditioning, was attracting interest among some horse trainers following its reported success in the dog training world. A clicker is a small plastic implement that makes a clear and precise "click" sound, and clicker training makes use of respondent conditioning to create an emotionally positive association between an environmental marker and a pleasant and biologically significant reward (a primary reinforcer), such as food or bodily scratches in the horse's case. Once this association has been created, by pairing the sound of the clicker with the presentation of food or scratches, the "click" can then be used to mark the behavior that is being trained, so that the trainer can then reward the horse for performing as required.

Left: *Clicker training, whereby a small plastic clicking tool is used by the trainer to create an emotionally positive association between an environmental marker and a pleasant reward such as food, has become a popular training method in the equestrian world.*

EQUINE ASSISTED THERAPY

The ability of horses to have a positive effect on humans' mental, emotional, and physical states has long been known. Their capacity to read and reflect back human emotion and their apparent honesty draw us to them in times of distress and horses are becoming well known for their ability to facilitate therapeutic change.

Horses have been used in physical therapy (or physiotherapy) since the early 1950s, letting disabled riders develop their motor skills in gentle and appropriate ways. Since then, the unique bond between human and horse has been incorporated into psychological therapies known variously as equine assisted therapy or equine facilitated therapy, and practiced in countries on every continent. Nontherapists also work with horses in areas such as coaching, business leadership, and personal development, and these activities come under the banner of equine assisted, or facilitated, learning.

Equine assisted therapy is characterized by being experiential, interspecies, nonverbal or verbal, and operating through a variety of models. The overarching field in which it sits, animal assisted therapy, was defined by Fine (2010) as "a goal-directed intervention in which an animal that meets specific criteria is an integral part of the treatment process." Two international organizations oversee the development of equine assisted therapy; the Federation of Horses in Education and Therapy International (HETI) and the Professional Association of Therapeutic Horsemanship International (PATH Intl.). EAGALA, the Equine Assisted Growth and Learning Association, takes a more operational role as the international organization that certifies equine and mental health professionals for careers in the equine therapy world. EAGALA has 4,500 members in 50 countries and has developed a therapeutic model held as the global standard for equine assisted psychotherapy and personal development and that includes a code of ethics upheld by a rigorous assurance process.

In equine assisted therapy, the client and a horse work together, along with a therapist, to promote emotional healing and growth. The therapy is used to help people with a variety of mental health issues, from addiction to low self-esteem, and, in particular, equine assisted psychotherapy has many potential benefits for the treatment of post-traumatic stress disorder (PTSD). Operation We Are Here in the United States and Help for Heroes in the UK are just two organizations that have information or resources on equine assisted therapy services. They support all those who have sustained injuries and illnesses during their service in the armed forces, changing and saving lives and helping those who have sacrificed and suffered for their country to recover and thrive.

To date, researchers have found no unified or empirically-supported theoretical framework that explains the therapeutic function of relationships between humans and horses. It has been suggested variously that:

• Horses possess intrinsic calming qualities while requiring in-the-moment attention.

• They offer opportunities for metaphor.

• Ethologically, as a prey species they must, for safety, attend to details that humans may not notice.

• Their size invites exploration of power and control issues.

• Their honest communication can reflect behavioral dissonance in the human.

• Human-horse interactions have anxiety-reducing effects on body arousal indicators.

Equine therapy helps to build rapport between therapist and client across a diversity of population groups, including psychoanalytic interactions in which patients revealed difficult thoughts by projection onto the equine assistant. The horse is a comforting or soothing figure that represents empathy and nonjudgmental positive regard.

The War-Horse ∽

The use of the horse in battle monumentally changed the way humans engaged in warfare.

At first, horses were not used in actual fighting but instead they allowed groups to raid further afield than they could by foot. Tribal raids began as early as 3000 BCE, in which groups of Eurasian steppe raiders would attack and loot villages and then retreat on horseback before the villagers could take action or retaliate. It is thought that numerous settlements in Eastern Europe were deserted because of these mounted raiders. However, this type of raiding did not spread far beyond the Eurasian steppes. Instead, a different kind of warfare with horses evolved: the chariot.

THE CHARIOT

The large chariots that came first were mostly just platforms that held a driver and an archer, and civilizations throughout Europe, Asia, and Northern Africa have evidence of horses and chariots starting around 2100 BCE. Wheels of a chariot were found in the Ural steppes and have been dated at between 2100 and 1700 BCE.

The invention of the spoked wheel replaced the solid wooden wheel and reduced a chariot's weight. This changed the nature of warfare for the civilizations of the ancient Near East, where the chariot became the essential resource in combat.

Below: *Detail from a painted casket discovered in the tomb of Egyptian pharaoh Tutankhamun depicts the king in battle against the Syrians in a horse-drawn chariot.*

Above: *Buffalo soldiers of the Tenth U.S. Cavalry at Fort Huachuca, Cochise County, Arizona ca. 1916. Cavalry warfare is suitable to almost any terrain and allows for fast raids, if necessary.*

Right: *The invention of the stirrup was hugely significant in the history of warfare. It enabled the rider to carry a usable weapon so horse and rider could charge at the enemy in battle.*

CAVALRY

The Iron Age in Mesopotamia saw the rise of the mounted cavalry as a tool of war. After 1000 BCE, cavalry consisted of a specialized force of mounted archers who had notable successes against tactics used by various invading equestrian nomads. Cavalry warfare is suitable to almost any terrain and eventually replaced the use of chariots, and because the horses of the Iron Age were still small at 13 to14 hands, they were successfully used as light cavalry for many centuries. The Mongols were famous for their horse-mounted archers and invaded and took over much of Asia using lightning fast raids in the thirteenth century.

For all of this time, stirrups had yet to be invented. Prior to the development of gunpowder, the stirrup was one of the most significant inventions in the history of warfare and they were quickly incorporated into many cultures' military equipment, such as the Byzantine Empire. By the end of the eighth century, stirrups began to be adopted in places in Western Europe, but not until Charlemagne's Empire were stirrups eventually distributed throughout Europe. Now, the horse was no longer just a ride to get the person into position, but could actually become part of the weapon. If the stirrup had not been invented, the lance and the arrival of the knight may have never happened.

CHANGES IN SIZE OF HORSE

Horses of a large size were first bred during Roman times in response to the demand of the legions for a strong cavalry horse, and during feudal times, the increasing weight of armor made the breeding of a large horse essential. The close of the twelfth century and the Crusades initiated greater use of European armor and this determined the development of large breeds of horse in the countries where knighthood flourished, such as Spain, France, and Flanders. The horses of Britain gained considerably in size when the Saxons and Danes imported larger breeds from the Continent. The development of the lance also occurred at this time.

With the rise in the use of gunpowder, knights began to decline and light cavalries again rose in prominence. Instead of the knight's heavy charger, breeding turned to a lighter horse bred for speed, staying power, and maneuverability. The conquest of the Americas happened largely because Americans on foot were no match for the Europeans on horseback. The horses that the Europeans brought to North and South America were used by the indigenous peoples to engage in warfare with the Europeans. Highly mobile horse regiments were also critical during the American Civil War.

WORLD WAR I

The most recent campaign in which horses were arguably involved before full mechanization, and on the widest scale, was World War I in 1914–18. A century on from this conflict, the horse's role has been brought back into sharp focus as people revisit the depravations and loss of this bitter struggle. Although horses were mainly used for transportation, both Germany and the UK also had a cavalry force of about 100,000 men each. At the beginning of World War I, there was a cavalry attack near Mons in Belgium, but it was one of the last of the war. Due to the rise of machine guns and trench warfare, the cavalry quickly lost favor as a form of attack. In March 1918, the British launched a cavalry charge at the Germans, and of the 150 horses used in the charge, only 4 survived.

Above: *During the nineteenth and twentieth centuries, while their role at the front line decreased, horses continued to be used for transporting and supplying armies.*

Military inventions at this time were still relatively new and were prone to breaking down so transport by horse was invaluable during the war. At the beginning of World War I, the British Army had only about 25,000 horses, so the War Office procured more from citizens to the extent that the British countryside was almost emptied of horses. They were transported to France and trained as cavalry horses or transportation horses. Horses were also heavily imported from the United States—at one point about 1,000 horses a day were loaded onto ships headed for Europe, and nearly one million American horses and mules served in the war. Finding enough food for the horses that traveled with the army was an issue that followed the armies throughout the whole war. By the end of the war, eight million horses died on all sides fighting in World War I, but two-and-half million were treated in veterinary hospitals and recovered.

MODERN-DAY USE

Horses still appear in modern military units around the world, but their purpose is now almost wholly ceremonial. These units retain close links with their former foundations, as in the UK's King's Troop Royal Horse Artillery Musical Ride, which uses original World War I guns and limber as part of their display. They demonstrate the military tactics used by the monkey men, who lie down with their horses on the ground, first with the horse resting on its breastbone so a soldier could use the horse as a shield and then completely flat on the ground.

These historic and ceremonial displays are repeated across Europe and farther afield, where the horse is still remembered by military organizations, including the USA's World War I Centennial Commission, which commemorated the contribution of the country's horses and mules to the war effort.

Below: *During World War I, the care of wounded war-horses was taken seriously, with two-and-a-half million surviving due to treatment in veterinary hospitals.*

The Sporting Horse ∾

Sport was not the first item on the agenda for the domesticated horse, but the use of the horse for warfare did lead to two of the earliest equestrian sports, namely racing and hunting. These were a natural consequence for those early nomadic tribes in Central Asia who used small and fleet-footed horses for their war games. Additionally, the game of polo also sits among those early equestrian sports.

RACING

Chariot racing predated mounted racing by several hundred years. The first recorded ridden race was at the Olympiad in Greece in 624 BCE. Racing became more formalized around the twelfth century in England, after crusading knights returned from their travels with Arabian horses, which were unlike the more solidly built animals that had to carry them and their armor into battle. As the Thoroughbred sport we know today, the formalization of racing began in the UK in the seventeenth century in the reign of King Charles II, with the first horse races being held at Newmarket, still recognized as the home of flat racing in the UK today. Known as the "Sport of Kings," the royal patronage that began the growth of the sport with King James I

developing Newmarket as the center of racing in its infancy continues to this day with Her Majesty the Queen of England a huge enthusiast and owner as was her late mother, The Queen Mother.

Below: *Her Majesty the Queen is a horse racing enthusiast and visits Ascot each year in June, when it hosts a week of special races known as Royal Ascot.*

HUNTING

Hunting from horseback was one of the earliest documented uses of the horse and dates back several thousand years BCE. In England, hunting is evident well before the Norman Conquest, although the invading French brought with them across the English Channel their own love of "la Chasse." Toward the end of the seventeenth century, the English began to hunt the fox, their quarry prior to this being the stag and the hare. It is illegal now in England to kill a fox with a pack of dogs and has been since 2005, but it is legal to use a pack to follow an artificial line, namely trail hunting or drag hunting.

POLO

The first polo matches were played in Persia (now Iran) more than two-and-a-half thousand years ago. Probably started by warfare between the tribes of Central Asia, the early army units quickly adopted it as part of their training methods, particularly for the King's elite troops. Royal patronage meant that polo was

practiced by kings, sultans, shahs, and caliphs and, therefore, it became known as the "Game of Kings." Never one to miss out on sporting activity, polo was adopted by British Army officers in the 1860s. An Olympic sport for a time, polo disappeared from the Olympic Games after 1936 but it still enjoys royal support in England, numbering both the Duke of Edinburgh and Prince Charles as former players, while Princes William and Harry continue to enjoy the sport

Above: *Illegal in England since 2005, fox hunting was once a staple pursuit of the wealthy.*

Below: *Known as the "Game of Kings," polo originally hails from Iran more than two-and-a-half thousand years ago. In 1860, the sport was adopted by the British Army and has always been a favorite with the Royal family.*

Above: *The different events that comprise rodeo are based on the working practices of cowboys as they herd cattle. The rodeo is designed to really put the skills of the rider to the test, and both agility and speed are crucial.*

RODEO

Rodeo is a series of competitive sports that arose out of the working practices of cattle herding in Spain, Mexico, Central America, South America, the United States, Canada, Australia, and New Zealand. The different events are based on the skills required of the working cowboys and vaqueros, and it has turned into a sporting event that involves horses and livestock, designed to test the skills of the riders.

The events include roping, steer wrestling, bronco riding, bull riding, and barrel racing, and they are governed by organizations that include the Professional Rodeo Cowboys Association. The first rodeo-like events began in the 1820s, following the American Civil War, when rodeo competitions emerged. The first was held in Cheyenne, Wyoming, in 1872, and rodeo continued to develop throughout the years, experiencing unprecedented growth in the 1970s.

DRESSAGE AND SHOW JUMPING

The more modern sports of dressage and show jumping really came to the fore when the horse ceased to have a role as a military and working animal. Classical horsemanship had always been present across the Western world, such as at the Cadre Noir in France and the Spanish Riding School of Vienna.

The sport of modern dressage as we know it today first appeared at the Olympic Games of 1912 and was designed to be a test of skill and obedience for the best trained military horses from the various European schools. It was not until the Los Angeles Olympics of 1932 that the more advanced movements of passage and piaffe were performed. As a sporting discipline, dressage was controlled by the British Horse Society (BHS), who also looked after horse trials (initially called combined training). British dressage as we know it today did not become a distinct entity until 1998, and certainly the UK lagged behind the continental Europeans in competitive dressage, some say because of their emphasis on hunting and eventing. It took a long time before the British caught up with their foreign neighbors who had forged ahead, breeding powerful and athletic sports horses. However, the UK is now firmly established as a nation at the forefront of the modern dressage sporting scene.

Jumping natural and rustic fences had been an activity engaged in for some time with horses, not least the obstacles encountered in the hunting field, and this began to form its own discipline in the shape of horse trials. Show jumping, however, initially evolved as a jumping test for hunters. The Royal Dublin Society staged what was called "a leaping competition" at the Dublin Show in 1865, and a year later a similar competition appeared at the Paris Show, although this time over rustic fences. Unlike dressage and horse trials, show jumping never fell within the auspices of the British Horse Society but was developed after World War II more seriously by the British Show Jumping Association, now rebranded as British Show Jumping.

The two world wars in the twentieth century considerably disrupted this process of sporting development. It was after World War II that the modern jumping seat of leaning forward also really developed; prior to that, most riders remained upright in the saddle or leaned backward, a tradition borrowed from the hunting field. The modern light seat that we now know was first used by Federico Caprilli (1868–1907). Caprilli studied horses loose jumping and photographed them to record the shape they made over fences, concluding that leaning back interfered with the horse's jumping movement and also restricted the horse by pulling on its mouth. He devised the light seat with a shorter stirrup so that the rider could remain in balance while the horse was in flight over a fence. Described as looking like "a monkey up a stick" in the early days, his style clearly caught on.

Below: *Dressage as a sport first appeared at the Olympic Games in 1912. It was designed to test the skill and obedience of military horses from various European schools.*

Centers of Breeding Worldwide ∞

The appearance of formalized centers for breeding horses across the world depended on several factors historically, including recognition of the value of the horse albeit for sport, agriculture, or war, and the presence of outside influences or interference.

EUROPE

In Europe, one of the oldest national studs can be found in Sweden, and this is Flyinge, which dates back to the twelfth century. Located near Lund in southern Sweden, Flyinge became established as a royal stud in 1661 by King Carl Gustaf X. Even now, still at the forefront of breeding the modern Swedish warmblood, a breed renowned as an excellent show jumper and dressage horse, the remit at Flyinge has expanded beyond just breeding. The stud supports other programs, such as a degree course in conjunction with the Swedish University of Agricultural Sciences. It is aimed at training future stable managers and riding instructors.

Mirrored across Europe, there are state studs in all the main European countries, most of which found their feet at around the same time during the seventeenth and eighteenth centuries; Dillenburg in

Below: Piber Federal Stud employees leading Lipizzaner colts from the Stubalpe Mountain to their winter stable, near Koflach, Styria, in Austria.

Germany, Pompadour in France, and the renowned stud at Piber in Austria, charged with the responsibility of maintaining the classical lines of the world famous Lipizzaner, used by the Spanish Riding School of Vienna. In addition to the state studs are numerous examples of private breeding enterprises, which are often larger and more financially successful. A good example is the Paul Schockemohle Service Station and Stud in Germany, run by the former German international show jumper. It is an incredible conglomeration of breeding expertise, top stallions, and world famous sports horse auctions.

Due to being geographically separate from the rest of Europe, much of the breeding of horses in the UK occurred on a piecemeal basis, with the exception of

racing. There are nine native breeds of pony, each with its own society fiercely proclaiming and defending the unique characteristics of that particular breed; however, there is no one overall native pony studbook that promotes the universal value and appeal of all these ponies. Compare that with mainland Europe: Germany, for example, has distinct warmblood breeds, such as the Hanoverian and Trakehner, both with their own studbooks, but closer examination of the development of these breeds reveals that they are an amalgam of other influences and bloodstock. This did happen to some extent with the native ponies of the UK, but much earlier in time and often accidentally, such as the shipwrecked vessels from the Spanish Armada running aground on the shores of Connemara in Ireland, from where Andalusian horses escaped and mixed with the native Connemara pony stock. Whereas modern warmblood breeders are comfortable looking away from their own breed if it improves the final outcome, this would not be permitted with the studbooks of the UK native ponies, which

have strict regulations about what can and cannot be registered.

The great contra to this in the UK is the Thoroughbred, the breeding of which is carefully controlled by Weatherbys' group of companies. Racing in the UK began to formalize its structure during the latter years of the eighteenth century, and alongside the development of established rules, Weatherbys began recording the details of the competing horses, their breeding, and results. This is now known as the *General Stud Book* (GSB) and this record has been maintained intact to this day, still in the care and control of Weatherbys. A new volume of the GSB is published every four years. Newmarket, described as the home of UK horse racing, became a focus for Thoroughbred breeders and still is even now with the presence of the National Stud and more than 60 other studs dedicated to the production of the Thoroughbred racehorse. Newmarket is also home to Tattersalls, Europe's leading Bloodstock Auction House, established in 1766 and holding nine sales per year in Newmarket and more at its other base in Ireland.

Above: *The Queen's stables, the Royal Stud at Home Farm on the Sandringham Estate, Norfolk, UK, with the statue of the racehorse Persimmon in the foreground. Persimmon was a British Thoroughbred whose racing career lasted from June 1895 to July 1897, during which time he ran nine times and won seven races.*

Above: *In the UK, the breeding of Thoroughbreds is stringently controlled by Weatherbys. During the late eighteenth century, it began recording details of racing Thoroughbreds.*

Right: *The free-roaming Dartmoor pony in England is one of the nine native breeds of ponies found in the UK. Each breed has its own society upholding and maintaining the unique characteristics of that particular breed.*

THE UNITED STATES

The situation in the United States in some ways is similar to that in the UK, with Kentucky being the central point for most horse production and certainly topping the list when it comes to Thoroughbreds. Theories abound on why Kentucky became the center for Thoroughbred breeding in the country, and certainly one plausible explanation is the ban imposed on betting on horse races, which took place toward the end of the nineteenth century. Kentucky was a state that did not subscribe to it and, consequently, horse racing took off there, attracting wealthy investors who wanted to bet as well as serious horse breeders who wanted to breed racing Thoroughbreds for them to bet on. Kentucky is, of course, also famous for its bluegrass, so called because in the early spring months, the buds have blue flowers that produce a canopy of blue-green. This grass is famed for its nutrient-rich properties, which is particularly high in calcium, so some say that this is why it is horse country; the truth, as ever, probably lies somewhere between the two.

Below: *Kentucky is the center of most horse breeding in the United States, and it is particularly famous for its Thoroughbreds.*

Right: *Shunsuke Yoshida at the award ceremony of the Challenge Cup, held at Hanshin racecourse in Hyogo, Japan.*

Bottom right: *There is a long tradition in the Middle East of breeding Arabian racing and endurance horses. As an extremely wealthy part of the world, it can support this tradition.*

JAPAN

Japan has something of a hidden Thoroughbred success story, behind which are the Yoshida brothers, with three stud farms and the Shadai stallion station to their name. The brothers are having an immense impact worldwide with their breeding program, having spent vast amounts of money investing in top-quality stock from Europe and the United States. Thoroughbred breeding is now big business, and with the ability to easily move horses around the world, there are not many parts of the globe out of reach.

THE MIDDLE EAST

The Middle East has a long and rich tradition of breeding fine Arabian racing and endurance horses, not least the distinction of providing the foundation of the English Thoroughbred. There is extensive royal patronage of both racing and endurance in the Middle East, and the country certainly has the wealth to support comprehensive and successful breeding programs. Some Arab sheikhs choose to have their breeding centers away from the Middle East, such as Darley, which is a global Thoroughbred breeding operation behind which stands HH Sheikh Mohammed Bin Rashid Al Maktoum, the ruler of Dubai. Darley stands stallions in Newmarket in England, in Ireland, in Kentucky in the United States, and on two stud farms in Australia. Darley also stands stallions in France and Japan.

Horses for Profit ∽

Below: *The entire history of the Western world and beyond has been shaped largely by human involvement with the equine. The exact value of the horse is immeasurable, but since its earliest domestication, it has contributed positively to the growth and success of humankind.*

Even before humans dealt in money and economics, horses have been contributing to our economy. Horse training as a profession appeared as early as 1400 BCE: Kikkuli from the land of Mitanni, which is now northern Syria and southeastern Turkey, wrote a book in which he describes a 184-day training cycle that begins in fall. This is the first known person whose profession was horses. Today, the horse contributes to many countries' economies.

THE ECONOMIC VALUE OF HORSES

Horses have always contributed to the growth and success of humankind since their earliest domestication. From providing a source of food to hundreds of years as a mount for transportation, a war-horse, a beast of burden, and an agricultural animal, the entire history of the Western world and beyond has been shaped by human involvement with the equine. As such a versatile creature, the value of the horse has been immeasurable, both in terms of social and historical development and in terms of its economic contribution. In post-industrialization countries, it is interesting to explore the contribution made by the

horse when it is no longer working but is largely used as a pleasure and sporting animal. The number of equines in the European Union lies somewhere around seven million. France comes out top of the member states with the highest number of horses at 840,000, closely followed by the United Kingdom. There is some debate over the final analysis because, despite strenuous efforts, there is still no centralized and enforced procedure for registering horses. So do horses currently contribute in any way to gross domestic product (GDP) now that their role is one principally of leisure and sport?

In the United Kingdom
The National Equine Survey established recently that the economic value of the British equestrian sector was growing and at £4.3 billion ($5.8 billion), spread across a wide and diverse range of goods and services. The report also estimated that there were 19 million equestrian consumers in Great Britain and that pleasure riding, which includes the sporting disciplines, was the most popular equestrian activity with 59 percent of that figure taking part in

unaffiliated events. In the British Horse Industry Confederation's Manifesto for the Horse, an equine sector economic contribution of £7 billion ($9.5 billion) was claimed, with a gross output of £3.8 billion ($5 billion) per annum. Racing made a huge contribution of £3.45 billion ($4.65 billion), not least through levies on betting and taxes amounting to £276 million ($373 million). Overseas trade was also strong at more than £500 million ($676 billion) and the UK equine industry averages an employment rate of 200,000 people, particularly in key rural areas, where employment opportunities are otherwise limited. Demographically, equestrianism was also found to engage a higher proportion of females, people with disabilities, and older people more than 45 years old, than many other sports.

In Europe

Within Europe, the equine industry remains one of the fastest growing economical areas. Sectors range from sport and agriculture to leisure and tourism. In terms of direct annual turnover, it is estimated that equestrian industries generate €1,100 to €2,800 ($1,300–$3360/£975–£2500) per horse. Equestrian industries are varied and include sport, tourism, agriculture, and breeding among others. There are also auxiliary industries that rely on equestrianism including veterinary, feed companies, and farriery.

Horse sport can be separated into two categories: racing and equestrian. There are an estimated seven million horses in Europe, with more than half this number involved in sport (as reported by the European Horse Network (EHN), October 2017). Equestrian events have a significant economic impact, being a global creator of jobs both directly (riders, grooms) and indirectly (farriers, feed and equipment companies). Horse racing is also a huge business and brings in approximately €6 billion (including revenue from betting) while providing 300,000 jobs (again, both directly and indirectly). Breeding of sports and race horses has a great economic impact with horses being exported around the world making up 32 percent of global sales. In the World Breeding Federation for Sports Horses 2011

Above: *In the United Kingdom, horse racing contributes, on average, a whopping £3.45 billion to the equine sector economic contribution, including levies on betting and taxes that amount to £276 million.*

Below: *Riding schools across the United Kingdom contribute to the equine sector economic turnover of about £7 billion per annum.*

Studbook rankings, European studbooks dominated, filling the top 23 places in Dressage, Eventing, and Show jumping. This demonstrates another successful growth area.

The equine industry plays a huge role within the agricultural sector, not only in terms of business and employment growth but also in promoting more ecologically balanced solutions, which provide a more sustainable growth for agriculture as a whole. Across Europe, there is a return to using the horse as a working animal on farms and also in forestry. As of June 2018, there are an estimated one million working equines, mainly within central and eastern EU countries. However, this growth is now spreading across western and northern parts of Europe.

Equestrian tourism has grown over the last 40 years and "has become one of the most important pillars of sustainable rural tourism" (EHN, 2018). The sector varies from carriage horses working in cities, such as Bruges, Vienna, and Prague, to horse-riding vacations. Horses are also making a comeback, serving communities by collecting waste and recycling as well as transporting children to school. In France the number of horses in this type of work has increased from less than 20 in 2001 to more than 300 reported in 2015. Not only does this increase jobs but it also reduces the impact on the environment, allowing for a more sustainable economic growth in this area.

The European equine industry remains an area of consistent economic growth and interest. However, further research is needed to address the ambiguity of this industry in individual countries.

Above: *Cart racing or harness racing in the United States is big business. As a whole, the American horse industry contributes almost $40 billion in direct economic impact to the national economy.*

Below: *There are an estimated seven million horses in Europe, and studies suggest that horse industries across the continent generate about €1,100 to €2,800 per horse.*

In the United States

The horse industry in the United States contributes $39 billion (£52.7 billion) in direct economic impact to its economy and supports 1.4 million full-time jobs. When indirect and induced spending are included, the industry's economic impact reaches $102 billion (£75.5 billion). It is a key contributor to the overall fabric of the U.S. economy. Horse owners and industry suppliers, racetracks, off-track betting operations, horse shows, and other industry segments all generate discrete economic activity contributing to the vibrancy of the overall industry. Of the total economic impacts reported, about $32 billion (£23.7 billion) is generated from the recreational segment; $28 billion (£20.7 billion) from the showing segment, and $26.1 billion (£1.9 billion) is generated from the racing segment.

There are an estimated 9.2 million horses in the country, with quarter horse and Thoroughbred being the two most popular breeds. Texas, California, and Florida are the leading horse states. Kentucky has long been known as a center of breeding and horse racing and is the home of the Kentucky Derby. It is estimated that the equine industry in Kentucky has an economic impact of almost $3 billion (£2.2 billion) and generated 40,665 jobs in 2012. The industry's tax contribution to Kentucky was about $134 million (£99 million).

In Australia

It is estimated that the horse industry in Australia contributes AU$6.2 billion (US$4.8 billion/£3.9 billion) per year. If you add in volunteer labor, that figure rises to AU$8 billion (US$6.2 billion/£4.6 billion) a year. Racing and associated activities contribute a little over half of the money, horse businesses, equestrian events, and breed events are also their own large industries.

In Canada

The Canadian horse industry contributes more than CA$19 billion (US$15 billion/£11 billion) annually. Farm activities with horses generate 76,000 full-time jobs. Activities with horses that take place off the farm, including racing, generate 9,000 full-time jobs. Overall, the Canadian horse industry supports more than 154,000 jobs. That is an average of one full-time job for every 6.25 horses in Canada.

Right: *Horse racing has always been big news in Australia. First held in 1861, the Melbourne Cup has a long prestigious tradition not just nationally but internationally. The race is known fondly as "the race that stops a nation."*

In New Zealand

The horse industry in New Zealand contributes an estimated NZ$1 billion (US$710 million/£526 million) to the economy, which is more than 0.5 percent of the GDP. It also directly sustains 12,000 full-time jobs. These estimates do not include the economic impact of the horse-racing industry, which is classified differently. Horse tourism has become popular in recent years, especially since *The Lord of the Rings* and *The Hobbit* trilogy of movies, which were all filmed in New Zealand (the native country of the movies' director, Peter Jackson). Much of the terrain where the movies were shot is inaccessible except by a horse, so horseback excursions are the ideal way to see this incredible landscape.

In the Middle East

The horse industry is also huge in the Middle East, with Dubai emerging as the focus of the horse business in the region. The Dubai World Cup race attracts an average of 65,000 spectators. Until 2017, the race offered the highest purse in the world at $10 million (£7.4 million). The horse tourism industry in Dubai is huge. Dubai is also famous for endurance races by Arabian horses. Other sports, such as show jumping, are gaining in popularity, too. The expansion of the industry is leading to an increase in homegrown businesses. Because of the climate, horses are more expensive to keep than in places such as the United States or Great Britain, but that does not seem to be swaying those who are trying to get into the growing industry in Dubai.

ECONOMIC IMPACT

Equine bodies within European countries and the United States are always eager to promote the equestrianism's economic impact on prosperity and the contribution it makes to economic success within society. There are other key benefits that surround this positioning of the equestrian industry and the importance of its contribution, and they are the benefits it brings to society as a sport and leisure activity, the employment opportunities that are created, and the importance of the welfare of the horse and continuing advances in veterinary treatment.

However, in less developed countries in the third world, the role of the horse is in stark contrast, remaining crucial to the survival of different peoples, because the horse is still a working animal. A Brooke Hospital report states that the rural parts of India still rely on more than one million horses, mules, and donkeys — described as "the invisible workers"— as beasts of burden and draft animals. They sustain the lives of countless people who, without them, would have no means of supporting themselves. They also provide critical links between different industries. It has proved difficult to quantify the economic impact of horses in the third world and one of the reasons for this is that they are not classified as an agricultural animal, because they do not produce food. The horse is still used as a food source, but this tends to be prevalent in the Western world, such as the United States and the newer members of the European Union, such as Romania and Hungary, who provide a horsemeat chain into central Europe, often with a significant welfare cost.

The main reason behind the Brooke's report is one of welfare, because that is the remit of the British organization. If the role of the working equids can be quantified somehow in terms of tangible economic benefit that undoubtedly derives from their use, then it will be possible to recognize and support these hidden animals. The World Organization for Animal Health (OIE) is currently developing the first Global Standards for the Welfare of Working Equids with the intention that governments engage with it and recognize the contribution and importance of these animals. There is an equine welfare issue at the heart of this, but a happy, healthy equine is able to work productively to sustain human life and survival, and so it remains in the human interest in the developing world to cherish and care for our equine companions.

Left top: Tourists embark on a horse-riding excursion on a farm in Glenorchy, New Zealand. This farm was used as one of the settings in The Lord of the Rings *movies.*

Left bottom: Polo is a popular emerging sport in the Middle East. Here, a match is being played on the open beach Jumeirah in Dubai, UAE.

Below: A shepherd with a caravan of mules carrying heavy supplies, food, and equipment in the Annapurna Base Camp in the Himalayan mountains in Nepal.

Ethology & Ethics ∽

The ethological study of horses is grounded, one would think, in scientific facts, given its perspective of examining naturally occurring behavior that enables the individual horse to survive and reproduce. However, the anthropocentric nature of human society guides humans to position themselves at the center of decision-making that affects the lives of animals, and this often involves making ethical judgments about how the horses in their care are managed and what interspecies activities are undertaken.

ETHICAL CONSTRUCTS OF THE HORSE–HUMAN RELATIONSHIP

The link between factual knowledge and ethical considerations is complex. For example, studying the consequences for equine welfare of various types of bridle makes the assumption that it is acceptable to use horses for humans to gain pleasure by riding them, as long as the individual horse's welfare is good. And assessments of welfare depend on what humans believe matters to the horse, whether that is to avoid discomfort, to experience positive emotions, or to live a natural life. To be able to make these assessments about welfare and well-being, a comprehensive understanding of the horse's natural ethology is essential, together with the ability to think about animal ethics in a reasoned way and not to rely on anthropomorphic feelings alone. Using feelings to justify our dealings with horses can make it difficult to explain to others why specific approaches may be beneficial or problematic.

Above: The well-being and welfare of a managed horse requires a thorough understanding of the horse's natural ethology and the ability to consider animal ethics in a reasoned way.

Right: Humans must take into consideration the environment a horse would experience in the wild. In this way, we can improve the quality of life of our equine companions.

TELOS

The notion of *telos* is a way of considering the quality of an animal's life by understanding the unique species-ness of that animal, the "horse-ness" of a horse in this case. The philosopher Aristotle held that animals have natures and functions that are founded in genetics and expressed both physically and psychologically. This is what determines how they live in their environments. Aristotle called this the *telos* of the animal. As a social animal, a horse needs to be with others of its kind; this is a species-specific need, as is the need to move and run as a prey species. The concept of *telos* relates to equine welfare problems, such as frustration and isolation, because horses are inherently beings that choose to live and interact in groups as part of their essential and genetically underpinned species-ness.

MORAL PHILOSOPHY

Human society has developed a belief that it has ethical responsibilities toward certain other species, including domestic animals that have contributed to human societies' economic development. Over many centuries, moral philosophers have developed theories about ethical positions, some of which might, in principle, underlie the various views about how it might, or might not be, acceptable to use horses for human enjoyment, emotional fulfillment, or economic gain.

Contractarianism

Contractarian morality takes the view that people are dependent on the respect and cooperation of others; humans are situated within a moral contract that animals do not share, thus any kind of animal use is permissible if it benefits humans. Animal protection is secondary and contingent upon whether people like animals; it also privileges certain species above others; in most English-speaking countries, for example, it is considered unacceptable to eat horsemeat, but it is culturally normal to consume other animals, such as cows or sheep.

ANTHROPOCENTRISM

Anthropocentric beliefs hold that humans are the most significant species on Earth and superior to all other beings. It places the human perspective as central and the most important, even when its effects will negatively impact the environment. Anthropocentrism is sometimes criticized by ethologists and environmental ethicists; however, others argue that humans have always used natural resources, including animals, for their benefit, and that anthropocentrism is therefore an adaptive feature of human behavior. Certainly, in terms of a contextual ethical approach to humans' duties toward animals, it can be argued that we have a responsibility to care for domesticated animals, such as horses, who, having been selectively bred for human enjoyment, may not be able to survive if returned to the wild, and that an anthropocentric approach would therefore follow, given the mental and physical resources culturally afforded to humans. However, in their own environment, an equine equivalent of water for a fish, horses would thrive and humans would not. So it is that humans have set up the world in their own species-specific needs, understandably and adaptively.

ANTHROPOMORPHISM

Animal caregivers are often accused of being anthropomorphic when they ascribe human traits, emotions, or behavior to animals. However, *critical anthropomorphism*, deriving from the fields of ethology and comparative psychology, refers to a perspective that utilizes the sentience of the human observer to generate hypotheses that reflect scientific knowledge of nonhuman species, their perceptual world, behavioral ecology, and evolution to generate new ideas that may be useful in gaining a more accurate and helpful view of how other species make meaning of their environments. For example, a horse owner may believe his or her horse likes to be wrapped up in a blanket in a warm stable, when reliable ethological and biological evidence shows that a horse is better adapted to live outdoors and typically without any additional clothing (subject to any medical or age-related considerations).

Anthropomorphism may also be seen when horse owners and caregivers attribute emotionally-charged words such as "boredom" to their horse, when there no evidence to suggest horses can actually experience these emotions.

Above: *Anthropomorphism is the attribution of human traits to an animal. The horse is no exception to this. The horse is far better suited to survive outdoors than most humans, so its need for outerwear is questionable, unless for medical reasons.*

Utilitarianism

In contrast to contractarianism, the utilitarian view foregrounds consequences as the only criterion. In 1789, Jeremy Bentham espoused maximizing pleasures and minimizing pains ("The question is not, Can they reason, nor Can they talk, but, Can they suffer?"), and in the 1980s animal ethics philosopher Peter Singer expressed this in terms of interests: If any being can suffer, then it has an interest in avoiding suffering; the interests of all beings should be taken into account equally, and the strongest interest should prevail no matter by which species this is held. In the example of meat consumption, while there may be arguments in favor of its reduction for reasons of its environmental effects, it would not, under utilitarian principles, be wrong to kill animals to eat, if each animal had lived a pleasant life, given the human pleasure that is gained from eating meat, which interest would arguably outweigh that of the animal's interest in continuing to live. The interesting question from a utilitarian perspective is, what is it about the horse that frequently places it in a different category of interest compared to meat production species?

Animal rights

The animal rights view resides separately in legal and moral perspectives; in addition, people now make meaning of this term in a political sense, too. In philosophical terms, moral rights are to be defended over and above claims that compete for any other reason. Animal rights philosopher Tom Regan argues that moral rights should be ascribed to all conscious creatures whose individual welfare is important to that being. They have the right to life and liberty and are not to be used for another's benefit, which illustrates the divergence from a utilitarian approach. Killing an animal deprives it of the good that would have come in the rest of its life and is a violation of its moral rights; in this context, killing any animal is morally unacceptable.

Respect for nature

Instead of being based on concern for suffering or a balance of interests, this viewpoint emphasizes protection from extinction and privileges natural species and genetic integrity. The value placed on animals is because they are a token of their species, so how do proponents of this approach perceive the domestic horse, selectively bred to be docile, easy to train, and dependent upon humans for survival, at the same time that suitable wild horse habitats are being eroded? Some environmental ethicists would regard it as less valuable, whereas others argue that wild or domestic is not such a clear-cut divide, as evidenced by rewilding breeding programs, such as the attempts to recreate the extinct tarpan, resulting in the Konik horse, Polish in origin and purported to contain a significant amount of tarpan DNA in its genome. This semiferal breed has been successfully released into nature reserves across Europe, although its genetic proximity to the tarpan is disputed.

Contextualism

Contextualism combines a number of different approaches to take into account broader considerations, such as the moral emotions of humans, whereby their own motivations of empathy and care

THE FIVE DOMAINS OF EQUINE WELFARE

In the 1960s, UK government guidelines for farm animals identified five basic freedoms that are the minimum that each animal should experience:

- Freedom from hunger and thirst
- Freedom from disease
- Freedom from excessive heat or cold
- Freedom of movement
- Freedom to act out normal behaviors

encourage them to bond with animals. Closely associated with moral relativism, this approach holds that the context in which an action is performed might determine whether the action is morally right, so, for example, a horse owner with a special commitment to their particular animal might view its ethical treatment differently in comparison to their view of the culling of free-roaming populations of horses, such as American mustangs or Australian brumbies.

Above: *The Konik horse is a semiferal breed of Polish origin. In an attempt to resurrect the extinct tarpan, the Konik has been breed and successfully released into nature reserves across most of Europe.*

CHAPTER 5

A Directory of Horse Breeds

Historic Breeds ❧

Most breeds of horse developed at a time when the horse was fundamental to many forms of human activity. Horses were a flexible organic power source (we still think of energy in units of horsepower), and they were integral to activities as diverse as travel and transport; warfare; heavy industry, including mining; long-distance communication, farming, and agriculture; sport; and hunting. The variety of breeds is a reflection of the differing demands placed on the horse in those varied fields as well as of the terrain and climates in the areas from which they originated. The physical and temperamental qualities required in a cavalry charger, for example, are different from those needed for a pit pony or a plow horse.

   Many of the horse's traditional functions have disappeared, mostly as a result of the technological revolution. No single invention has done more to speed the decline of the horse than the internal combustion engine; throughout the developed world, the car drove the horse out of cities. In New York in 1900, the horse population is estimated to have stood at 100,000 (a single bus required the pulling power of twelve horses a day); by 1912, there were more motor vehicles on the streets of Manhattan than there were

horses. And in the short 20 years between the first and second world wars, most cavalrymen became tank crews and drivers of armored vehicles, relegating the horse to transporting supplies and troops where needed.

Above: *Prior to the invention of the internal combustion engine, the horse was fundamental to people's ability to get from A to B other than on foot.*

LOST FOREVER

Once the practical reason for their existence evaporated, breed populations naturally began to shrink, and many breeds disappeared altogether. In the British Isles, six breeds became extinct between 1900 and 1973, including the Goonhilly and the Galloway. Across Europe, breeds such as the German Emscherbrucher became extinct while others, such as the Augeron from France, were amalgamated into other breed standards. Similar attrition occurred in the United States, despite the fact that horses remained a practical mode of transport across the country's vast prairies and on ranches. Among the breeds lost was the Narragansett Pacer. It was the first breed to be developed in the country—George Washington owned one—but it disappeared before the twentieth century. The Abaco Barb, believed to be descended from horses that were shipwrecked in the Bahamas during the Spanish colonization of the Americas, is another notable extinction. Despite efforts to maintain a small surviving herd, the last mare died in 2015.

As well as the many recorded breeds known to be extinct, many undocumented breeds have probably been lost globally. Because data collection is variable and often inadequate, especially in parts of the world where horse breeding is less formalized, it is not possible to have a complete picture of breeds that are currently at risk. In 2007, the UN's Food and Agriculture Organization published the State of the World's Animal Genetic Resources report, which classifies around one in five horse breeds as endangered due to declining population size.

SOME OF THE MOST WELL-KNOWN AND CHARACTERISTIC HISTORIC BREEDS

1. Arabian	6. Exmoor	10. Highland Pony
2. Andalusian	7. Camargue	11. Icelandic Horse
3. Thoroughbred	8. Friesian	12. New Forest Pony
4. Fell Ponies	9. Haflinger	13. Norwegian Fjord
5. Trakehner		14. Ardennais

Above: *Now unfortunately extinct, the Narragansett Pacer was one of the first breeds to be developed in the United States during the eighteenth century. It does live on, however, in the Tennessee Walking Horse, because the Pacer was regularly crossed with other breeds.*

A THREATENED EXISTENCE

Europe

Of the existing 14 horse breeds native to the United Kingdom, 12 are under threat. Breeds in a critical state include the Suffolk Punch, the Cleveland Bay horse, the Eriskay Pony, the Dales Pony, and the Hackney Horse and Pony. Elsewhere in Europe, the Poitou Donkey of France is endangered. Austria's Lipizzaner, famed for the elegant dressage steps performed at Vienna's Spanish School of Riding, is endangered, but it is being maintained by an active conservation program.

North America

The American Cream Draft is the only draft horse breed to have been developed in the United States. Now rare, it is considered critically endangered. The horses that make up the Colonial Spanish group of breeds (or strains) are descended from Spanish stock brought to the Americas from the Iberian peninsula (Spain and Portugal) by the conquistadors. Those most at risk include the Banker Pony, the Carolina Marsh Tacky, and the Florida Cracker.

The Canadian, or Cheval Canadien, known as the "little iron horse," is also considered critical. The breed descends from stock sent to Canada in the 1660s by King Louis XIV of France. Many were exported from Canada to the United States in the 1860s for use as war mounts during the Civil War. By the 1880s, their numbers were dwindling. They recovered in the twentieth century, but they are now in danger once again.

Clockwise from above: Breeds in danger of disappearing altogether include the American Cream Draft (above), the South African Nooitgedachter pony (above right), the Japanese Kiso (below right), and the Austrian Lipizzaner (below).

Middle East and Asia

The Caspian was believed to be extinct until it was rediscovered in Iran in the 1960s. It is an ancient breed, thought to date back 3,000 years, and it is also the smallest in the world. Breeding programs in the United States, United Kingdom, Australia, New Zealand, and Scandinavia have helped to expand the population, but it is now deemed critically endangered.

Asian breeds that are endangered include the beautiful Akhal-Teke of Turkmenistan, which is thought to predate the Arabian; it owes its shimmering golden coat to a hair structure that is unique. In Japan, most of the native breeds are critically endangered, including the Kiso, and in India the Deccani is almost extinct.

Africa

Endangered breeds considered to have been developed in Africa include the English Halbblut horse, which is considered critical, and the Nooitgedachter pony. The feral Namib, which roam the Namib Desert, is also under threat.

CONSERVATION AIMS

Preserving domestic animal breeds, including horses, is now recognized as an important part of conserving global diversity. As well as being a significant part of our cultural history, at-risk breeds may well have unique characteristics that will be useful in withstanding new challenges brought by climate change and new livestock diseases in the future. Within breeds, too, the broadest possible genetic diversity also needs to be maintained. Interbreeding due to low stock or popular breed lines can undermine this, but modern technology has allowed for the collection and storage of semen and embryos, which can help to preserve the integrity of the breed.

The FAO's 2007 Global Plan of Action for Animal Genetic Resources, adopted by 109 countries, aims to preserve the genetic heritage of horses and other domestic animals and ensure that they remain available to future generations. National organizations, such as the UK's Rare Breeds Survival Trust, The Livestock Conservancy in the USA, Heritage Breeds Canada, and the Rare Breeds Trust of Australia, are working to identify the causes of declining breed populations and to help stabilize them. These organizations support a network of rare breed associations and registries, which play an important role in monitoring and maintaining the quality of specific breeds.

From Historic to Modern Breeds ∽

Przewalski's horse

It is probably a well-accepted premise that horse breeding changed across the globe in radical terms when the horse was no longer used in warfare and postmechanization, when it was no longer required for agriculture. At that point, some breeds disappeared and breeders' preferences tended to move toward a lighter, more athletic animal that was suitable for leisure and for sport. This has had a huge impact over the direction of horse breeding during the past 100 years, with many breeds evolving into something altogether different, while others have stood still, surviving, but with no current purpose other than the enjoyment of enthusiasts and the stated aim of preserving the breed.

There are three subtypes and three prototypes that developed from the ancient breeds into the modern horse as we know it today. The three subtypes are the domesticated horse, the Przewalski's horse, which is a breed that has never been domesticated, and the tarpan, which is now extinct.

Andalusian

THE THREE PROTOTYPES OF THE ANCIENT HORSE ARE:

(a) **the warmblood** subspecies or now extinct Forest horse, thought to have contributed to the development of the warmblood breeds of northern Europe as well as older heavier horses, such as the Ardennais.

(b) **the draft** subspecies, a small sturdy heavyset animal with a heavy hair coat, which arose in north Europe and adapted to cold, damp climates. This prototype is probably related to today's draft horses and even the Shetland Pony.

(c) **the oriental** subspecies, a tall, slim, refined, and agile animal that arose in Western Asia. It adapted to hot, dry climates well. It is thought to be the ancestor of the modern Arabian horse and Akhal-Teke.

EUROPE

The British Isles are unusual in that there are nine native breeds of ponies, which is not found elsewhere in the world. Scotland has the Shetland Pony and Highland Pony, and Ireland has the Connemara. Wales has four Welsh ponies: the Welsh Mountain Pony, section A; the Welsh Pony, section B; the Welsh Pony of Cob type, section C; and the Welsh Cob, section D. England has the two northern breeds, the Dales Pony and Fell Pony; the southern New Forest Pony; and the southwest Exmoor Pony and Dartmoor

Pony. These ponies owe their heritage to the social and economic history of these islands, woven into the fabric of the development of the countries. Native ponies are popular for riding and driving, showing, and jumping, and they make a great first cross with the Thoroughbred to produce a small and athletic sport horse, although several of the undersupported breeds are being watched by the Rare Breeds Survival Trust (RBST), because numbers are low.

Along with many other nations, the United Kingdom had its fair share of heavy horses, because people once relied greatly on these powerful animals as beasts of burden and as draft horses. Notable among these are the Shire, the northern Clydesdale, the Percheron (originally a French breed), and the Suffolk Punch, which is distinct among these as being clean of limb—in other words, having no feathers so it could work effectively in the heavy clay soil of East Anglia. Of course, the United Kingdom also boasted the Thoroughbred, which by the mid-twentieth century already had several hundred years of evolution as a sporting animal, and it was to this that the Europeans turned during the twentieth century, especially after World War II.

Mainland Europe has no history of ponies, although arguably there are some smaller horse breeds, such as the white horses of the Camargue. Europe has its own draft horses, such as the Italian heavy draft, the Ardennais, and the Jutland from Denmark. After the war and with the mechanization of agriculture, some of these breeds have been retained by enthusiasts. They have largely begun to fall out of favor, although the thriving continental meat trade provided an outlet for some and still does; this is a mainland European factor that is not seen in the United Kingdom.

Additionally, there were warmblood horses, which although not as heavy as the draft breeds, needed some improvement as riding horses. After World War II, the continental Europeans began to improve these breeds and they began to steal the show from the British in producing modern show jumpers and dressage horses. Thoroughbred blood was used to lighten the frame and conformation of Hanoverians, Trakehners, Belgian warmbloods, Danish warmbloods, and the French Selle Français. The old chunky, square-topped warmbloods with large and often plain heads, which were derided by the British during the 1960s and 1970s, have changed beyond recognition into lighter framed, athletic, and beautiful sport horses.

The continental Europeans have long clustered their breeding into large, well-organized operations, from covering stallions to auctions, and so the modern warmblood now dominates both the international dressage and show jumping scene. The British has followed this trend to some extent, but warmblood breeding in the United Kingdom is necessarily more piecemeal. Warmbloods are being favored over the more traditional English sporting cross, which would be the Irish Draft crossed with a Thoroughbred. Since the loss of the three additional endurance phases in top level horse trials—the two roads and track phases and the steeplechase—there has also been a tendency to favor warmblood breeding in three-day eventing, because there is such a premium now on the dressage and show jumping elements. However, a recent reversal to more old-fashioned, big galloping cross-country tracks has wrong footed this move, because the only animal that has both the endurance and the stamina, and that can really perform effectively over these courses, is the three-quarter or seven-eighths Thoroughbred, the remaining element of the cross is usually a heavier breed, such as the Irish Draft, which provides the power.

Interestingly, Spain and Portugal did not join the European trend seen in France, Germany, and Denmark of favoring big athletic warmbloods. Instead, they worked with their own native Iberian breeds of the Andalusian and the Lusitano, smaller-framed horses that have a strong working tradition that includes, of course, the bullring. They lend themselves greatly to the more collected work of advanced dressage, but they were never competitive in the modern sense until Spain burst onto the international dressage scene with the charismatic

Icelandic

Rafael Soto Andrade claiming silver at the 2004 Athens Olympics. His passion and flare totally exploded the serious domination of the Dutch and Germans, with a refreshing Latin flavor and a completely different type of horse. Since then, the Iberian horses have found favor with many other riders, proving that it is possible to be successful in dressage without a European warmblood.

Although there are not really any pony breeds on mainland Europe, Scandinavia bucks this trend completely, because the topography and climate lends itself to the development of smaller, hardy, surefooted animals that act both as draft and ridden animals, strong enough to carry an adult rider. Among these are the classic Norwegian Fjord Horse and the Icelandic Horse, which—although standing no higher than 13.2 hands (54 inches/137 cm)—is never referred to as a pony by the Icelanders.

THE UNITED STATES

After becoming extinct in the Americas between 8,000 and 12,000 years ago, the horse was reintroduced to these continents in the sixteenth century when the Spanish arrived, bringing horses with them, and other European settlers moved into the New World. Many horses escaped and some were left behind when settlements were abandoned. These horses quickly spread from South America to North America. Some were domesticated by the indigenous peoples of the arca, but others became feral horses and established territories and breeds that still exist today. One of those breeds is the Colonial Spanish Mustang. Mustangs, as they are commonly called, are best known for roaming the Western United States. There used to be hundreds of wild horses roaming all over the Old West, but as the population of humans increased, the population of mustangs decreased. In 1971, the Wild Free-Roaming Horse and Burro Act was passed to protect mustangs from being slaughtered. The Bureau of Land Management now manages the population of these horses.

There is still a population of horses that exists today that are true Colonial Spanish Mustangs—its DNA has been traced back to the horses originally brought over from Spain. These horses live on the islands of the Outer Banks in North Carolina, which is where their ancestors originally landed when the ships transporting them were shipwrecked and they swam to shore. In the 1920s, there were 5,000 to 6,000 horses roaming the islands, but now there are just a few hundred.

The range of modern breeds in the United States is fairly diverse. The Rocky Mountain horse was developed in eastern Kentucky in the 1800s. It was bred as a versatile farm horse with an easy four-beat gait that made riding and traveling long distances easy and comfortable.

The most popular and populous breed in the United States is the American Quarter

Appaloosa

Horse. The breed originated in the seventeenth century. It was first bred in Virginia and the Carolinas. These horses are known for being an extremely versatile creature and the fastest horse over short distances. The Appaloosa breed became well known through its use by the Nez Perce. The breed is known by four characteristics: coat pattern, mottled skin, white sclera, and stripped hooves.

The American Bashkir Curly is a breed descended from a herd of three horses found by the Damele family roaming in Nevada in the late 1800s. No one knows the origins of these horses. Their defining feature is an extremely curly coat that covers their entire body, their mane and tail, and even inside their ears. As the horses age, many of the curls disappear. Their hair can also be used for spinning.

The Pinto breed is defined by its color. It originated with Native Americans and was later adopted by Western settlers. There are two types of patterns: tobiano, which is a white horse with large patches of color on the body, and overo, which is a color horse with white markings on its side.

The Nakota breed is one of the last descendants of the wild horses of North Dakota. The breed was almost eradicated until a group was preserved in the Theodore Roosevelt National Park. They are a sound horse with great stamina. They tend to mature slowly, but they develop strong bonds with humans and are reliable horses. The Missouri Fox Trotter breed began to develop in the Ozark mountains when settlers arrived in the 1820s. The eponymous gait developed as a result of the hilly terrain of this area. It has become a popular mount of long distance trail riding

THE MIDDLE EAST

Diversity continues farther afield with the Arabian horse, which is hugely prominent in the Gulf States due to the patronage and wealth of some of the royal families who support the growing sport of endurance. Seemingly across the world, countries make use of their equine heritage and adapt their native breeds to modern purposes.

INDIA

The Kathiawari is a desert breed of horse that originated in Western India. It has tipped-in ears, which is characteristic of a couple of breeds that developed in India. It is used as a utility horse for driving and riding. It is also used for mountain police terrain.

RUSSIA

In Russia, the Budenny breed was established after World War I. The Russian Revolution left the horse population extremely depleted. Today, they excel in eventing and other sports. They are most often chestnut with some white markings. The Orlov Trotter is a breed that was developed in Russia. These horses were traditionally used by Russian nobility for riding and harness racing. Today, the breed is preserved in Russia and Ukraine.

SOUTH AMERICA

The Criollo horse—the native horse of Uruguay, Argentina, Brazil, and Paraguay—originated directly from a shipment of 100 Andalusian stallions in 1535. This breed is known for its long-distance endurance, which is linked to a low basal metabolism, and its hardiness. It can survive extreme cold and heat, and it can live on limited water and dry grass. The Paso Fino breed was developed from horses brought to the Dominican Republic by Christopher Columbus to be used as mounts. The unusual smooth gait that replaces the traditional trot is the breed's hallmark. The breed has flourished in Puerto Rico and Colombia. The national horse of Brazil is the Mangalarga. It has developed mainly from breeds that originated in Spain. The breed has been unchanged since the 1800s. The Chilean Horse (Corralero) is the oldest registered native American breed, the oldest registered stock horse breed in the western hemisphere, and the oldest registered breed in South America. It has a reputation for courage, stamina, surefootedness, even temperament, and trainability.

Welsh

HEIGHT RANGE
11 to 15 hands
(44 to 60 in./112–152 cm)

COUNTRY/REGION OF ORIGIN Wales

COMMON COLORS

gray	bay
palomino	liver chestnut
chestnut	buckskin
chestnut roan	strawberry roan

One of the most popular and charismatic of the nine native UK breeds, the Welsh breed is divided into four sections, the first two of which are ponies. The Welsh section A is the smallest, standing at no higher than 12 hands (48 inches/122 cm) and is described as the Welsh Mountain Pony, often with a beautiful dished face that betrays the Arab influence from earlier centuries. The Welsh section B, the Welsh Pony, is larger and can go up to 13.2 hands (54 inches/136 cm); it retains the fine chiseled features of the section A, but it is a larger-frame pony and described by the breed society as a riding pony. Both section A and section B are easy to identify; they move well, with section B having more free-flowing gaits than the smaller pony, which tends to be shorter in the back and show more knee action. They are characterized by their verve and panache, and they can easily be seen in terms of color. Grays abound but also palominos, roans, chestnuts, and bays, with plenty of white—socks, stockings, and blazes—due to the sabino gene.

The section C, the Welsh Pony of Cob type, is a true blend of the Welsh Pony but with Cob blood; it should stand no higher than 13.2 hands (54 inches/136 cm). The section D, the Welsh Cob, is one of the most popular sections; from 13.2 hands (54 inches/136 cm), it has no upper height limit and is renowned as a quality ride and drive animal.

Dales

HEIGHT RANGE
14 to 14.2 hands
(56 to 57 in./142–144 cm)

COUNTRY/REGION OF ORIGIN England

COMMON COLORS

- ■ black
- ■ brown
- □ gray
- ■ bay
- ▨ roan (occasionally)

Hailing from the north of England, the Dales Pony is a heavily built large-frame pony, originating from the Upper Dales of Tyne, Allen, Wear, and Tees in North Yorkshire. Mainly black or dark brown in color, although occasionally a gray or bay may be seen, the usual height is anywhere from 14 to 14.2 hands (56 to 57 inches/142–144 cm). Strong limbs with a good length of neck combine with a short coupled back to provide an excellent riding or driving animal, while still retaining a neat pony head.

This breed of pony originally worked in the mining areas of the north of England, hauling lead, copper, and iron ore from the mines. More recently, they were used in the coalfields, underground if there was enough depth, but more commonly above ground to pull wagons. Despite their evident strength, they are free-moving ponies, popular today for riding and driving. They still retain a native surefootedness and soundness derived from centuries of work on harsh terrain.

Connemara

HEIGHT RANGE
12.2 to 14.2 hands
(50 to 57 in./127–144 cm)

COUNTRY/REGION OF ORIGIN Republic of Ireland

COMMON COLORS

- ■ black
- □ gray
- ■ brown
- ■ dun

- ■ bay
- ■ occasionally roan and chestnut

From the south of Ireland, the area of its namesake, the Connemara is a quality native riding pony. The breed may stand anywhere up to 14.2 hands (57 inches/144 cm) and is a lighter-frame pony than some of the other natives, but it still shows plenty of bone and strength. The Connemara has a free-flowing action that lends itself to ridden work, without the higher knee action of other pony breeds. A true pony head gives way to a good length of neck and a back that is slightly longer. It is most often seen in a gray coat color, but duns are also commonplace. Legend states that the quality in the pony came from Arabs and Andalusian horses that were brought ashore from Spanish ships in the fifteenth century, and there is certainly an athleticism about the pony combined with its native ruggedness.

Originally used as both pack animals and riding ponies, because they were large enough to carry a man, the modern Connemara pony is a popular choice as a family pony or sport pony. It combines the freedom of movement and an excellent jump, with a pony toughness and intelligence, while still retaining all its hardy native characteristics.

Dartmoor Pony

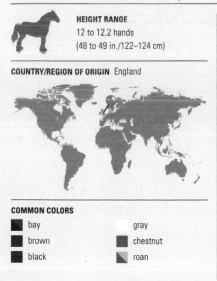

HEIGHT RANGE
12 to 12.2 hands
(48 to 49 in./122–124 cm)

COUNTRY/REGION OF ORIGIN England

COMMON COLORS
- bay
- brown
- black
- gray
- chestnut
- roan

The Dartmoor Pony is a small pony with the desired height not exceeding 12.2 hands (49 inches/127 cm). Hailing from Dartmoor in the West Country region of England, the Dartmoor Pony should not be confused with the Exmoor Pony, because they are different ponies. The Dartmoor Pony's color range can be anything from bay to chestnut with the rare, occasional gray—white markings are permitted but not in excessive amounts whereas the Exmoor Pony is a universal color of dark bay or brown with no white markings permitted. The Dartmoor Pony has a finer head than the Exmoor and an average length of both neck and back, so it is proportional.

Both the Dartmoor Pony and Exmoor Pony were originally employed as pack animals: the Dartmoor to carry tin from the moors and the Exmoor as a sometime pit pony. Both were used extensively within the farming community as beasts of burden. Today, these ponies are favored for riding and showing, and they make an excellent child's pony.

New Forest Pony

HEIGHT RANGE
up to 14.2 hands
(up to 57 in./144 cm)

COUNTRY/REGION OF ORIGIN England

COMMON COLORS
any color permitted except pinto

One of the largest of the UK's nine native pony breeds, the New Forest Pony is completely associated with the forest in which it still roams. The pony is a strong working type, but it has always had enough quality to make it a riding pony instead of just a pack animal. The upper height limit is 14.2 hands (57 inches/144 cm) and most New Forest ponies are around this height, although they may be smaller and breed classes are often split between smaller and larger height ponies. They retain a strong pony appearance, but evidence of early breeding with Spanish and Arab influence—there was a royal stud at Lyndhurst in the fifteenth century with such stallions—has resulted in an animal with quality and athleticism. Colors vary but are usually darker bay, brown, or chestnut. A small amount of white on the head and lower limbs is permitted.

As well as being a working animal, the speed and athleticism of the pony means that it has long been favored as a riding pony, too. There is plenty of historical evidence of the ponies being raced in the nineteenth century for sport. Today, they are popular as mounts for both children and adults, having enough quality of movement to make an excellent sport pony while retaining all the hardiness of a native pony, a winning combination.

Highland Pony

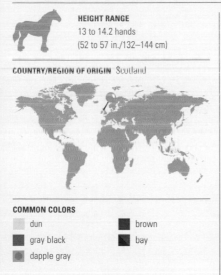

HEIGHT RANGE
13 to 14.2 hands
(52 to 57 in./132–144 cm)

COUNTRY/REGION OF ORIGIN Scotland

COMMON COLORS

- dun
- gray black
- dapple gray
- brown
- bay

One of three Scottish native breeds, the Highland Pony is a strong, square pony with a frame and size that is completely proportional to its height. No bigger than 14.2 hands (57 inches/144 cm), the pony may be smaller as long as the conformation still retains the element of "squareness" and proportionality, so the neck and back are relative in length to one another and the length of leg. Despite its strength and build, the Highland Pony is still very much a pony in terms of appearance and type, retaining a pony head with that natural inquisitive pony look. Most commonly gray or dun, the breed society defines an extensive range of dun shades from mouse to silver, all of which should exhibit the distinctive eel or dorsal stripe—a black line—down the center of the back, and sometimes "zebra stripes" on the limbs and shoulders, too.

Originally used as a pack horse and as both a ride and drive animal, this pony is capable of carrying the weight of a man. Due to its immense strength and versatility, it worked as a plow animal on Scottish farms and is still used today to haul stags off the hills. Its derivation from such a harsh climate makes it an extremely hardy pony with exceptional feet. As a ridden mount, it is perhaps seen more commonly under saddle with adults instead of children.

Shetland Pony

HEIGHT RANGE
7 to 10.2 hands
(28 to 41 in./71–104 cm)

COUNTRY/REGION OF ORIGIN Scotland

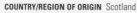

COMMON COLORS

■ bay	☐ gray
■ black	◣ piebald
■ brown	◺ skewbald
■ chestnut	☐ dun

The smallest and most diminutive of the nine native UK breeds, the Shetland Pony is unmistakable both in terms of its height and type. An ancient breed from the islands north of Scotland, its location has ensured that the pony has remained relatively untouched by outside breeding influences. It is thought that the pony's size was accounted for by the spartan surroundings and the lack of forage; only small animals could survive in these conditions. The maximum height for adult ponies is about 10.2 hands (41 inches/104 cm) although they may be smaller, and the Shetland Pony Stud-Book Society now has a registration category for miniature Shetlands. Colors can be varied, including skewbalds and piebalds, the only one of the nine native breeds to permit this. The darker solid colors tend to feature little if any white.

For centuries, the Shetland Pony was used for all kinds of work around the islands other than as a driving pony, because there were no roads. Because no location in Shetland is more than 4 miles (7 km) from the sea, they were involved in the fishing industry, with hair from their tails converted into fishing line. In more recent times, they were used on the mainland as pit ponies, the only breed small enough to access the mines.

Shire

HEIGHT RANGE
16.2 to 18.2 hands
(65 to 73 in./165–185 cm)

COUNTRY/REGION OF ORIGIN England

COMMON COLORS

■ black
■ brown
■ dapple gray
■ bay
□ gray

The Shire is the largest of the UK's draft horses and is distinctive not just for its height but also its trademark of four white socks with immense flowing feathers. The Shire horse can stand anywhere from 16.2 hands up to 18.2 hands (65 to 73 inches/165–185 cm), and most stallions are at least 17 hands (68 inches/173 cm). Usual colors are bay, brown, and black, although there are also grays, the dark colors will have plenty of white below the knee and hock. The head of the Shire is noble, with a slight Roman nose and a high set, and it has an arched neck that leads to a long sloping shoulder and powerful, short coupled back, usually with immense depth in the body.

With such size and power, the Shire horse has featured predominantly throughout all human activities for the last 1,000 years, as a horse of war and agriculture to the time of mechanization, where it worked alongside the combustion engine as a draft horse. In dangerous decline, these horses are still valued for the breed alone and shown in hand; there is even a growing trend to ride them now to help boost numbers and, most recently, a Shire racing series has been developed at some UK racecourses.

Clydesdale

HEIGHT RANGE
16 to 18 hands
(64 to 72 in./163–183 cm)

COUNTRY/REGION OF ORIGIN Scotland

COMMON COLORS

■ bay
■ black
■ chestnut
◨ bay roan
□ gray

A somewhat newer breed than the Shire horse, the Clydesdale first came to prominence in Lanarkshire in Scotland in the eighteenth century. Clydesdale was the old name for Lanarkshire, hence the name of the breed, and it is sometimes referred to as "the Scottish Heavy horse," because of this geography. Flemish stallions were used to upgrade native stock into the animal that we see today. As a horse distinct from the Shire, the Clydesdale has, if anything, a more Romanesque nose and often has extensive white markings on all four legs, but specifically on the hind legs, higher than the hock, and white splashing on the underside of the body. Typically standing anywhere around 17 to 18 hands (68 to 72 inches/173–183 cm), like the Shire, the predominant colors are often dark but, unlike the Shire, it is possible to also see roans. The Clydesdale's confirmation is similar to that of the Shire; it is the head and color that usually tells them apart.

With dwindling numbers, the Clydesdale faces similar challenges to the Shire horse. Quietly used as a trekking horse under saddle in the north of England for a number of years, the Clydesdale is now joining forces with the Shire and appearing under saddle.

Suffolk Punch

HEIGHT RANGE
15 to 17 hands
(60 to 68 in./152–173 cm)

COUNTRY/REGION OF ORIGIN England

COMMON COLORS

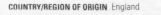

- liver chesnut
- dark chesnut
- red chesnut
- light chesnut
- bright chesnut

The Suffolk Punch is distinct from the other heavy breeds, not only because of its color but also because of the lack of feathering on the leg. Bred in Suffolk to work the clay lands of East Anglia, the lack of feather was deemed a necessity in the heavy, wet soil. A smaller breed than the Shire and the Clydesdale, ranging from 15 to 17 hands (60 to 68 inches/152–173 cm), the Suffolk Punch is nonetheless incredibly powerful and is used throughout agriculture and forestry as a draft animal. It is always chestnut, but note that the spelling of this color actually has the letter "t" omitted, so it is spelled as "chesnut." Different shades of chesnut are accepted; originally there were seven variants, but the modern breed society now recognizes five. The Suffolk has a deep and powerful neck tapering to the shoulder and a long and graceful back. The legs are short in proportion to the body, which is where the immense power comes from.

Another heavy breed with dangerously low numbers, much is made of the breed in its native county, and it is still used for forestry work and within agriculture. It is included within the remit of the recently formed British Ridden Heavy Horse Society, and there are now Suffolk Punch horses appearing under saddle alongside others of the UK's heavy breeds.

Percheron

HEIGHT RANGE
16.2 to 18.2 hands
(66 to 73 in./165–185 cm)

COUNTRY/REGION OF ORIGIN France

COMMON COLORS
gray
black

The Percheron is originally a French breed, and it is unclear when it arrived in the United Kingdom. Some say with William the Conqueror, but there is certainly evidence of the importation of heavy horses exhibiting Percheron breeding, ironically from North America, during the later years of the nineteenth century. The Percheron and its influence may be found all around the world, wherever French settlers left their mark, although the British Percheron Horse Society, which was formed in 1918, has set itself apart and devised its own breed standard. Imported animals may be accepted if they meet the relevant criteria.

The Percheron is distinctive in that it is always gray or black and with relatively little feather, so it is easy to identify. Standing at a minimum height of 16.2 hands (66 inches/ 165 cm), the Percheron is not as large as the Shire, but it is nonetheless a strong animal but with perhaps the best frame for a riding horse out of all the UK heavy horses. If you look at its history across the globe, where it was certainly used as a mounted animal in warfare, it is possible to see these antecedents in the breed's current conformation. Supported now by the British Ridden Heavy Horse Society, the Percheron is now part of a growing movement to show these horses under saddle, in part to give them a useful role and thereby to support their numbers.

Camargue

HEIGHT RANGE
13.2 to 14.2 hands
(53 to 57 in./134–144 cm)

COUNTRY/REGION OF ORIGIN France

COMMON COLORS
gray

Known as the white horses of the Camargue, because of their color, these iconic animals populate the area of this name in southern France. The horses have been present in this landscape since time immemorial. They are a tough and hardy creature, a pedigree developed in the marshes and wetlands, with harsh winters followed by exceptionally hot summers. Always gray in color, with the exception of the foals, which are born a dark color, the horses are in reality the size of a pony, standing at 13.2 to 14.2 hands (53 to 57 inches/134–144 cm). They have, however, a distinct short horse head. It was not until relatively recent times, in 1976, that the French government sought to introduce some regulation to the breed and registered all Camargue breeders, setting a breed standard to protect this iconic horse.

The horses are still employed within the Camargue by the "Gardians," so-called "Camargue Cowboys," who use them to round up the cattle. The region celebrates the white horses in festivals and processions, and they are used as riding and driving ponies for adults and children alike. The Camargue horse shows a remarkable aptitude for all equestrian disciplines, including horse ball and more recently, long-distance riding for which its surefootedness and stamina make it ideally suited.

Pottok

HEIGHT RANGE
11 to 14.2 hands
(44 to 57 in./112–144 cm)

COUNTRY/REGION OF ORIGIN Basque Country

COMMON COLORS

- bay
- black
- chestnut
- piebald
- skewbald

The Pottok is a small pony breed native to the Pyrenees within the Basque Country, so is a shared inhabitant of both France and Spain. This challenging mountainous environment has created a tough and surefooted pony considered by the Basque people to be an iconic emblem of this geographical region. The Pottok is deemed to be a breed at least several thousand years in age. It has remained historically isolated from other equine populations due to its location, including within the Basque country—modern genetic testing has revealed some stark differences between ponies in the northern Basque Country and the southern Basque Country. Some scientists view them as distinct animals in that they are different breeds. A census that took place in 1970 established about 2,000 purebred Pottoks to the south of the Pyrenees and 3,500 to the north, which rang alarm bells, because these are smaller numbers than anticipated.

Unlike similar breeds of pony, the Pottock has short but slim legs. The Pottock has had an interesting range of uses throughout its history. Due to its location and coloring, it was a popular choice for smugglers moving contraband back and forth across the Pyrenees. The Pottock was also used as a pit pony in both France and the United Kingdom; crossbreeding to produce piebald Pottock ponies was a popular choice for many years in the circus.

Ardennais

HEIGHT RANGE
15 to 16 hands
(60 to 64 in./152–163 cm)

COUNTRY/REGION OF ORIGIN France

COMMON COLORS

- bay roan
- iron gray
- dark chestnut
- liver chestnut
- bay
- brown
- palomino

The Ardennais, sometimes referred to as the Ardennes, draws its name from the area in which it originated, shared between Belgium, Luxembourg, and France. The Ardennais is possibly one of the heaviest of all the working horses, with an incredible musculature, perhaps as a result of the desired lower height compared to some other draft breeds. Belgian Draft blood was used in the nineteenth century to increase the heaviness of the breed. Most commonly bay or roan, other colors, such as gray and even palomino, are permitted. Their conformation is square and stocky, with short limbs and a short neck with vast depth. Their incredible power and equable temperament made them ideal for a variety of work premechanization and the advent of modern warfare. From the early years of the twentieth century, all of the three countries began their own studbooks for this breed, although there is much pooling of bloodlines between them.

The continental appetite for horsemeat means that many of the breed are destined for the table, so numbers remain healthy for this reason, if nothing else. More recently, there has been a move to use these heavy horses in forestry work, for leisure purposes, and in tourism, so they may be seen again in some of their former working roles.

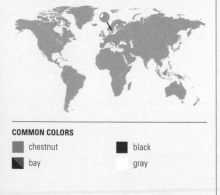

Jutland

HEIGHT RANGE
15 to 16.2 hands
(60 to 65 in./152–165 cm)

COUNTRY/REGION OF ORIGIN Denmark

COMMON COLORS

▉ chestnut	▉ black
◣ bay	☐ gray

Jutland is a peninsula that is shared between continental Denmark and the north of Germany. The breed to which it gives its name is a heavy breed, most commonly found today as a chestnut horse, although other colors used to be commonplace, such as bay, gray, black, and roan. The first studbook for Jutlands was opened in 1881, which was not long after the importation of foreign blood to improve native stock, with Suffolks and Ardennes being among these imports. Chestnut has now been identified as the breed color and this is due in part to the influence of the Suffolk Punch.

A team of chestnut Jutlands pull the wagons for the famous Carlsberg brewery around Copenhagen and have achieved something of a cult status as a result. Carlsberg promotes the breed as much as they promote their beer. In real terms, the breed nowadays has purely a show purpose, as with other heavy breeds, and numbers have dwindled down to around 1,000 horses compared with a healthy population of approximately 15,000 after World War II, with the resulting mechanization of agriculture taking its toll, as it did with so many other heavy breeds.

Hanoverian

HEIGHT RANGE
15.3 to 17.2 hands
(61 to 69 in /155–175 cm)

COUNTRY/REGION OF ORIGIN Germany

COMMON COLORS

- gray
- chestnut
- black
- brown
- bay

There is no doubt that the modern Hanoverian horse is the prime example of selective breeding, resulting in a beautiful and athletic sport horse. Controlled and careful processes have produced an animal that is in great demand for its athletic abilities, and this is why there are close to 20,000 registered Hanoverians in the main studbook. It all began with a state-owned stable of stallions in 1735, followed by the actual formation of the breed studbook in 1888.

The modern Hanoverian differs considerably from the postwar warmblood horse, because demand at that time was for a strong horse that could be used on the farm, as a carriage horse, and in the arena of war. As tastes changed, the old square-topped Hanoverian has developed into a quality, lighter-frame animal with exceptional gaits and movement. The controlled introduction of foreign bloodlines, for example, the Thoroughbred—a policy not followed with other continental warmbloods—plus rigorous mare, stallion, and foal inspections and gradings, have meant that only the best stock has been promoted, a breeding policy that has clearly paid off. Today, Hanoverians feature at the highest level of international competition as dressage horses and show jumpers.

Trakehner

HEIGHT RANGE
15.2 to 17 hands
(61 to 68 in./154–173 cm)

COUNTRY/REGION OF ORIGIN Germany

COMMON COLORS

- bay
- gray
- chestnut
- black
- brown

Named after the town of Trakehnen, which was formerly in East Prussia, a state stud for the breed was established there in 1731 and remained in this location until 1944, when East Prussia disappeared into Russia and Poland. Unlike other continental warmbloods whose breeders involve outside bloodlines if they are good and fit the desired criteria for improvement to existing stock, the Trakehner Studbook has always been largely closed, with only a few outside Arab and Thoroughbred bloodlines permitted. The Trakehner is considered to be perhaps the lightest framed of all the warmblood breeds and the one most closely allied to the Thoroughbred. This is because Thoroughbred blood has been influential in developing the quality of the breed, a trait also reflected in the breed's temperament. The Trakehner has therefore often been used by other warmblood breeds to provide an injection of quality. The Trakehner Verband, like its continental neighbors, also exhibits and grades its stock.

The Trakehner is a top-quality, light, modern sport horse beloved of dressage riders and show jumpers. It has a beautiful frame that produces cadenced and elegant movement and is highly sought after as a sport horse across the disciplines.

Belgian Warmblood

HEIGHT RANGE
16 to 17 hands
(64 to 68 in./163–173 cm)

COUNTRY/REGION OF ORIGIN Belgium

COMMON COLORS
- chestnut
- bay
- brown
- black

Late starters in the warmblood breeding stakes, due in no small part to the Belgian government's concern about protecting the dwindling population of native Brabant horses, otherwise known as the Belgian Draft, Belgian Warmblood breeding did not really begin to take serious shape until the second half of the twentieth century. The Belgian Warmblood Studbook (BWP) was founded in 1955. Without their own native stock of riding horses or heavier warmbloods to use, the Belgians imported different breeds from France, Germany, and the Netherlands to create a modern riding horse. There are many different influences within the breed and the horse can vary to some degree in both temperament and conformation, depending on which bloodlines it has come from.

The Belgians follow the same rigorous inspections of stock of other European warmblood breeds but, interestingly, there is no specific breed standard as such, just a uniformity of purpose. If an animal is inspected and falls within those criteria—soundness, good movement, health—then they are graded accordingly. Belgian Warmbloods are regularly found at the top end of international competition in both dressage and show jumping.

Danish Warmblood

HEIGHT RANGE
15.3 to 17 hands
(61 to 68 in./155–173 cm)

COUNTRY/REGION OF ORIGIN Denmark

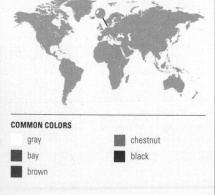

COMMON COLORS

◻ gray
◼ bay
◼ brown
◼ chestnut
◼ black

This is another continental warmblood that began to flourish after the end of World War II, when Danish breeders covered their own native stock, such as the dwindling Fredericksborgers, with other European warmblood breeds, principally the Trakehener and the Thoroughbred. The result is a stunning modern sport horse with a beautiful light frame reminiscent of the Thoroughbred, but with bone and definition that provides the athletic movement so beloved by dressage riders and show jumpers. Despite the evident amount of Thoroughbred blood, both directly and indirectly via the Trakehner, the Danish Warmblood has a pretty level temperament.

Danish Warmbloods are graded and shown, with elite auctions for showing the best stock. Because of the level of Thoroughbred influence, the Danish Warmblood has also been used for eventing, because it has the quality of gaits to excel in the dressage and show-jumping disciplines and the speed and stamina of the Thoroughbred to run successfully across country.

Selle Français

HEIGHT RANGE
15.3 to 17.3 hands
(61 in. to 69 in./155–176 cm)

COUNTRY/REGION OF ORIGIN France

COMMON COLORS
- ■ bay
- ■ chestnut

The Selle Français is the modern French sport horse, and, like the Belgian Warmblood, it can show a wide range of variation within the breed. The French used Thoroughbred blood to cross with their own existing Norman stock in the nineteenth century, and a studbook appeared in the late 1950s. In 1965, when Algeria became independent from France, the French National Stud closed the Barb Studbook and transferred the best mares into the Selle Français Studbook. Stock inspections and grading of both native Selle Français animals and any others wanting to interbreed are held to ensure the best possible breeding outcomes.

The modern Selle Français is a popular sport horse and has excelled in all the modern equestrian disciplines to international level. Its color is usually bay or chestnut. Although there is no physical universal breed standard, instead there is a uniformity of purpose set against certain criteria, such as soundness, athleticism, movement, and conformation. Temperament can vary because of the varying breeding influences that have been used to create the modern horse and, to some degree, type as well.

Thoroughbred

HEIGHT RANGE
15.2 to 17 hands
(61 to 68 in./154–173 cm)

COUNTRY/REGION OF ORIGIN England

COMMON COLORS

■ brown	■ chestnut
◤ bay	□ gray
■ dark bay	

The Thoroughbred horse was bred in England, originally to race, and it still fulfills this role today and has subsequently gone on to be exported to all corners of the globe. It is, without exception, the most dominant influence on sport-horse breeding in the world, claiming the title of being the fastest of all horses over typical race-length distances. It has had a huge impact on improving many different types of breeds, and it also provides a first cross for heavier animals to create a powerful and athletic sport horse.

The modern Thoroughbred can trace its lineage back to the three founding fathers of the breed—the Byerley Turk, the Darley Arabian, and the Godolphin Arabian, which were present in England throughout a period spanning the seventeenth and eighteenth centuries. Darley and Godolphin are two names that are still today synonymous with modern horse racing. The Thoroughbred is a light-frame horse with tremendous quality of bone. It is designed to go quickly with a long neck that is low set on the shoulder, making it ideal for galloping, and sloping, powerful hindquarters and long cannon bones. For speed and athleticism, the Thoroughbred has no equal. It has influenced many modern breeds; for example, there is a large percentage of Thoroughbred blood in the modern continental warmblood.

Irish Draft

HEIGHT RANGE
15.2 to 16.3 hands
(61 to 65 in./154–166 cm)

COUNTRY/REGION OF ORIGIN Ireland

COMMON COLORS

- bay
- brown
- gray
- dapple gray
- chestnut
- black
- dun

The Irish Draft horse is one of Ireland's most famous exports. Although it began life as a working horse, it was never designed to be used solely for draft work and is not defined as a heavy breed; however, it undoubtedly has huge power. It was developed as a multipurpose animal, so it is a horse that is strong enough for agricultural work yet it can also be ridden and driven.

The Irish Draft developed in part from the Irish Hobby, which was a small and agile horse, now extinct, but believed to be influential in the origins of the Connemara pony. There is a shared lineage between the Irish Draft and the Connemara via the Irish Hobby, with a sprinkling of Connemara somewhere back in Irish Draft ancestry. Because heavier horses became less popular after World War II, the Irish Draft was found to be an excellent first cross with the English Thoroughbred to produce what is now described as an Irish Sport horse. The Irish Ministry of Agriculture began a studbook in 1917 but, unfortunately, all records were destroyed in a fire in 1922. Post-World War II, numbers declined rapidly, and in 1976 the Irish Draught Horse Society was founded to save the breed. Numbers are still at a low level, but they are being monitored by the Rare Breeds Survival Trust.

Andalusian

HEIGHT RANGE
14.3 to 15.2 hands
(579 to 61 in./145–154 cm)

COUNTRY/REGION OF ORIGIN Spain

COMMON COLORS

■ bay

□ gray

■ black

■ dapple gray

Hailing from the south of Spain, this ancient and charismatic breed of horse has always been used as a riding horse, principally in the bullring, where it was required to show speed, bravery, athleticism, and huge trainability, qualities that it has in abundance. The Andalusian was highly prized because of its stockmanship abilities and speed. It was employed in warfare for these reasons, and its grace and nobility meant that it was valued, not just because of its amazing qualities, but also because of its appearance.

A small compact horse, the Andalusian is most commonly gray and has a wavy mane and tail that are usually kept long. It has a distinctive head with a slightly concave profile—one might say Romanesque, but it differs to the Roman nose found in other breeds, because the head is shorter and of more quality. The neck is graceful and sloping, proportional to the length of back, and the quarters rounded and powerful, which make this horse ideally suited to the collected movements of high school work, echoing its earlier antecedents in the bullring. The Andalusian shot to fame when Rafael Soto Andrade propelled the Spanish dressage team to a silver-medal winning position at the Athens Olympics in 2004 with his charismatic gray horse Invasor.

Lusitano

HEIGHT RANGE
15 to 15.3 hands
(60 to 61 in./152–155 cm)

COUNTRY/REGION OF ORIGIN Portugal

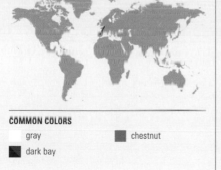

COMMON COLORS

■ gray
■ dark bay
■ chestnut

The Iberian neighbor to the Andalusian horse, these two breeds share many similarities and a similar heritage. The Lusitano Breed Society claims that it is the oldest ridden horse in the world. Evidence has been found on the Iberian Peninsula, now shared between Portugal and Spain, which includes archeological finds of bones and skulls, plus engravings and cave paintings, some of which have been confirmed as dating to around 20,000 BCE. Prized as a mount for warriors and invaders over the centuries, the Lusitano did not actually become a distinct breed with its own studbook until 1967. Prior to this, horses from this part of Spain and Portugal were called Iberian horses or Andalusians. The Lusitano, so called because this is the ancient Roman name for Portugal, is perhaps a stronger and stockier version of the Andalusian, in some part due to the fact that Portugal still practices mounted bullfighting. A need to use the Andalusian for other purposes, such as equitation and dressage, has resulted in a lighter, more elegant horse with some Arab influence. However, this is not to say that the Lusitano lacks quality, but it has a muscular force and power more synonymous with its role in the bullring and as a stock horse.

Fjord

HEIGHT RANGE
13.2 to 14.2 hands
(53 to 57 in./134–144 cm)

COUNTRY/REGION OF ORIGIN Norway

COMMON COLORS

■ brown dun	■ red dun
gray	white dun
■ yellow dun	

The Fjord pony, indigenous to Norway, has been in existence for at least 4,000 years. It has had little other breeding influences, partly due to its location in the mountainous regions of western Norway and the other key factor, the climate in Norway. Arab and Thoroughbred blood would not have crossed well with a horse that has to survive in such harsh conditions. The Fjord remains, therefore, one of the purest of equine breeds.

Known as a horse, the Fjord actually stands at only about 14.2 hands (57 inches/144 cm), so it is pony height. However, this strong, stocky, and square animal is easily able to carry a man's weight. Its color makes the Fjord instantly recognizable; 90 percent of all Fjords are a brown dun and the remaining 10 percent, red dun, white dun, yellow dun, or gray. One of their most striking features is the mane, the outer hair of which is white while the inner hair is black. Breeders and enthusiasts usually keep the mane short so that it stands upright, revealing the two-tone color palette, which is somewhat a hallmark. The dun Fjords also carry dorsal or eel stripes—the black stripe along the back—and zebra markings on the legs. Like a miniature draft horse, the Fjord is used under saddle and makes an excellent driving animal, being strong enough for farm and forestry work and surefooted enough for mountainous terrain—a true utility horse.

Icelandic Horse

HEIGHT RANGE
13 to 14 hands
(52 to 56 in./132–142 cm)

COUNTRY/REGION OF ORIGIN Iceland

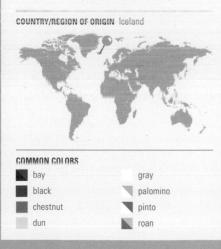

COMMON COLORS

- ■ bay
- ■ black
- ■ chestnut
- ☐ dun
- ☐ gray
- ◩ palomino
- ◩ pinto
- ◪ roan

Similar in some respects to the Fjord of Norway, the Icelandic Horse is in reality the size of a pony, but it is strong and powerful enough to carry a completely grown man and survive in harsh conditions. It was brought to Iceland by Norse settlers in the ninth and tenth centuries. No other breed of horse has been allowed into Iceland for the last 1,000 years, so the Icelandic Horse perhaps claims the crown over the Fjord as having the least external breed influences of any equine breed. The Icelandic Horse has proved popular worldwide, with nineteen different breed federations, but any horse that leaves Iceland is not allowed to return. Iceland started its own breed registry in 1904.

A tough and hardy animal, the Icelandic Horse has a double layered coat to protect against the harsh elements. It is unique in that it is five gaited. In addition to walk, trot, canter, and gallop, the Icelandic horse has a fifth gear called tölt; this is a four-beat gait with the same footfalls as the walk, but it can be performed at different speeds—it can be potentially fast and quickly cover ground. It is estimated that there are 80,000 Icelandic Horses in Iceland, which is remarkable considering the human population is just over 300,000. The Icelandic people are proud of their horses and, as well as riding and racing them, they are used extensively in the tourism industry.

Arab

HEIGHT RANGE
14 to 15.3 hands
(56 to 61 in./142–155 cm)

COUNTRY/REGION OF ORIGIN Arabian peninsula

COMMON COLORS

gray

bay

chestnut

black

Along with the English Thoroughbred, the Arab horse must be one of the most influential breeds across the globe. It is certainly one of the oldest. Originally a desert horse bred by the Bedouins in the Middle East, the Arab has always been prized as a creature of immense beauty, speed, and stamina. The Arab is one of the most handsome of horses, with an elegant and sculptured head, often with ears that slightly curve inward toward one another. Its chiseled head leads to a long and graceful neck and a body that is proportionate, with a high set on tail often carried out and up at an angle, a hallmark of the breed.

The Arab horse, due to its value and appearance, has no history as a working horse—it was almost always a saddle horse—but it was used as a cavalry horse due to its turn of speed. In recent times, the Arab has tended to become most associated with the sports of racing and endurance riding; this is not because it is not suited to other disciplines but simply because other breeds—often, ironically, with earlier Arab influence—have become more specifically bred for the modern sporting disciplines of dressage, show jumping, and three-day eventing.

German Riding Pony

HEIGHT RANGE
13.2 to 14.2 hands
(53 to 57 in./134 144 cm)

COUNTRY/REGION OF ORIGIN Germany

COMMON COLORS

■ bay	■ gray
■ chestnut	◧ palomino
■ brown	□ dun
■ black	◩ roan

The German Riding Pony began to develop after World War II, when the role of all horses started to change and the horse became increasingly viewed as a leisure animal. Like its counterpart, the British Riding Pony, the ideal stamp was deemed to be a well conformed sport horse in miniature, with the necessary athleticism and gaits to be a successful competition horse, while retaining desirable pony characteristics, such as temperament and pony charm. Initial attempts by the Germans were not successful—they used Arab and Thoroughbred bloodlines on Fjord and Haflinger ponies. Moving farther afield, they turned their attention to the Welsh breed, already infused with a lot of quality Thoroughbred and Arab blood, and found that their breeding programs were more successful in producing the kind of pony they wanted.

The modern German Riding Pony is like a warmblood sport horse in miniature, and today's breeders stick to specific bloodlines to create this stock. As with all German breeding programs, ponies must pass rigorous inspection if they are to go on to be used for breeding purposes.

Akhal-Teke

HEIGHT RANGE
14.2 to 16 hands
(57 to 64 in./144–163 cm)

COUNTRY/REGION OF ORIGIN Turkmenistan, Central Asia

COMMON COLORS

- golden buckskin
- palomino
- cremello
- perlino
- gray
- chestnut
- bay
- black

The Akhal-Teke, which originates from Turkmenistan, is one of the most distinctive of all breeds. It is notable for its shiny, metallic coat, which makes it probably unique among horse breeds and easy to identify. Its unusual sheen comes from hair that is so fine it is almost like silk.

The Russians opened the first studbook for the breed in 1941 after the country and the horse had been subsumed by the Russian Empire. Today, there are about 6,000 of these horses worldwide. It is a larger horse and appears similar to the English Thoroughbred, although the head is different, almost Roman at the top and across the forehead before tapering to a finer muzzle. The back is long in proportion to the animal. The Akhal-Teke was used for racing as its conformation dictates, and it is still popular for both racing and endurance riding. It has been debated whether the Byerley Turk, one of the founding stallions of the English Thoroughbred, was in fact an Akhal-Teke. Palomino and buckskin are common colors and the breed is also characterized by long ears and almond-shape or hooded eyes. The Akhal-Teke is the national emblem of Turkmenistan; the breed appears on the country's stamps and banknotes, as well as its coat of arms.

Jeju

HEIGHT RANGE
11 to 12.3 hands
(44 to 49 in./112–125 cm)

COUNTRY/REGION OF ORIGIN South Korea

COMMON COLORS

- black
- brown
- chestnut
- bay
- cream
- skewbald
- roan

The Jeju horse hails from an island off the Korean Peninsula, from which it takes its name. The Jeju is closely linked to the ancient Mongolian horse, and there is plenty of historical evidence of interbreeding. During the thirteenth century, the Mongols controlled Jeju, bringing with them their own horses that interbred with the native island ponies. Small in stature, the horse is, in reality, pony size; it is sometimes called, "Gwahama," which means "short enough to go under a fruit tree." A healthy population of around 20,000 Jeju horses dwindled to just more than 2,000 by the early 1980s, and the South Korean government stepped in to give the breed National Monument status. Following this, the Korean Racing Association built a large racecourse south of Jeju si, where registered Jeju horses are raced. Horseback tourism is also a huge feature of island life and a thriving industry, and there is an annual Jeju Horse Festival held in October. Paradoxically, horsemeat is a prized delicacy on the island and is served in several restaurants in Jeju, both cooked and raw.

The Jeju horse is a substantial animal for its height, being both strong and gentle. It has a square and relatively long body compared to the length of the leg, hence its power and ease in working on the land, and its ability to carry an adult under saddle.

American Quarter Horse

HEIGHT RANGE
14 to 16 hands
(56 to 64 in./142–163 cm)

COUNTRY/REGION OF ORIGIN United States

COMMON COLORS

- sorrel
- bay
- black
- brown
- buckskin
- palomino
- gray
- dun

The Quarter Horse, or American Quarter Horse as it is commonly known, is the most popular breed of horse in the United States. Its conformation and function is inextricably linked with the development of the country through its value as a stock horse; as the mount of the cowboy; in rodeo and other equestrian disciplines associated with western riding, such as cutting and barrel racing, and even as a racehorse. These are all qualities that mean that it is an attractive option for those who still want a stock horse, but it also has a role as a sport horse in the modern equestrian disciplines.

The American Quarter Horse is the ultimate equine athlete, which is why its appeal has endured, from the requirements of the Wild West to the modern day. Early Spanish influence from the conquistadors brought Arab and Barb bloodlines to native American horses, followed later by the Thoroughbred, which came with the English colonists; the American Quarter Horse owes much of its athletic prowess and speed to these early ancestors. This popular breed falls into two distinct types: the small, stockier, well-muscled horse, which is favored for western sport, and the taller, lighter-frame horse, which is used for racing. With a Registry of around three million horses, the American Quarter Horse Association is currently the largest breed society in the world.

American Paint Horse

HEIGHT RANGE
14.2 to 16.2 hands
(57 to 65 in./144–165 cm)

COUNTRY/REGION OF ORIGIN United States

COMMON COLORS white base color combined with:

- bay
- black
- brown
- sorrel
- chestnut

Evolving from American Quarter Horse and Thoroughbred bloodlines, the American Paint Horse is classified as an actual breed in its own right, not a painted type of another breed of American horse. The key fact with any paint horse is the coloring, but beyond that he specific bloodlines and conformation that ensure that the American Paint Horse is categorized as a distinct breed.

The American Paint Horse Association, second in size in the United States only to the American Quarter Horse Association, registers horses as tobiano, overo, tovero, or solid. Solid coat colors are also eligible for registration in the same way that the Appaloosa breed accepts solid colored horses that are not spotted. However, the American Paint Horse Association also stipulates conformation and breeding that must be American Quarter Horse, Thoroughbred, or preexisting registered paint horse. This stands the breed apart from the Pinto, which is a broader concept also based on color but one permitting a lot of different breed types. Some paint horses carry the Sabino gene also found in Welsh ponies and may, therefore, exhibit blue eyes and roan spots.

Appaloosa

HEIGHT RANGE
14 to 16 hands
(56 to 64 in./142–163 cm)

COUNTRY/REGION OF ORIGIN United States

COMMON COLORS

◉ blanket spot	■ chestnut
● leopard spot	◢ palomino
◣ bay	◩ buckskin
■ black	cremello

The Appaloosa, known to all as the spotted horse, is an American breed with ancient origins; cave paintings depict illustrations of spotted horses from prehistory. The spotting is created by overlaying certain identified patterns on top of solid base coat colors, and this is one of the reasons why the Appaloosa historically and genetically has many different equine breed influences. The name, Appaloosa, is thought to hail from the Palouse River, which is in the northwest area of the United States occupied by the Nez Perce peoples. It is thought settlers referred to the horse as the "Palouse Horse" and this name eventually evolved into Appaloosa. During the 1877 Nez Perce War, the Nez Perce lost most of their horses in the conflict and the breed declined considerably; only a few hardy breeders persevered. In 1938, the Appaloosa Horse Club was formed and the breed has since gone from strength to strength; it is one of the most popular breeds in the country.

Common breed influences in the modern Appaloosa are Thoroughbred, Arab, and American Quarter Horse, producing a hardy, quick, and versatile animal that is popular in Western riding. However, the Appaloosa also has good enough athletic ability to make it a common choice for other sporting disciplines.

Boerperd

HEIGHT RANGE
14 to 16 hands
(56 to 64 in./142–163 cm)

COUNTRY/REGION OF ORIGIN South Africa

COMMON COLORS

- ■ black
- □ gray
- ◣ palomino
- ▨ dun
- ■ chestnut
- ◣ pinto

The modern Boerperd is a South African breed derived from the Boer Horse, a much older breed. Although currently low in numbers in its native country, the modern Boerperd is a strong, athletic, and muscular horse exhibiting the influence of a number of foreign breeds. These were exported to the area by settlers in earlier times, namely the Flemish horse, Hackney, and even Cleveland Bay.

During the Boer War, the numbers of Boer Horses were dramatically reduced by the conflict; horses were lost in warfare and the British followed a policy of destroying the horses whenever they came across them to deny the Boer access to their mounts. At the end of the war, in 1902, efforts began to register the remaining horses, although this lost momentum after World War I. After World War II, a formal breeders association began in 1948, and a second registry was opened in the seventies. Both still exist today and have established strict criteria and selection for the registration of stock. The modern Boerperd is recognized as a universal sport horse. Registered horses are described as having five gaits and two different types of action are permitted: a lower knee action or a higher knee action, which is a nod perhaps to some Hackney ancestry.

Kabarda

HEIGHT RANGE
14.1 to 15.1 hands
(56 to 60 in./143–153 cm)

COUNTRY/REGION OF ORIGIN Russia

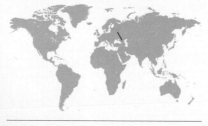

COMMON COLORS
- gray
- black
- bay

A horse of the Caucasus and named after the Kabardino-Balkaria region of Russia, the Kabarda (also known as the Kabardin) is a strong and muscular horse with proportionate conformation, well defined limbs, a deep chest, quality head, and a coat color that can be only black, gray, or bay. The Kabarda has evolved to survive in challenging conditions and combines hardiness with quality. Unusually, the blood of the Kabarda has a heightened oxidizing ability, essential for work at altitude.

The Kabarda has a history going back at least four hundred years. It was bred from varying horses of the Steppes by mountain tribesmen, who needed a creature that was surefooted and could endure both the harsh terrain and the climate. After the Russian Revolution, the Kabarda dwindled, and in an effort to boost numbers and introduce fresh bloodlines, the Kabarda was crossed with the Thoroughbred to produce the Anglo-Kabarda, which still exists today. The Kabarda itself is divided within the breed into three types: the Oriental (due to its quality derived from Arab influence); the Original (more closely allied to the foundation horse, it is stockier and well muscled); and the Ordinary (larger and used formerly as a carriage horse and for light draft work).

Dosanko

HEIGHT RANGE
13 to 13.2 hands
(52 to 53 in./132–134 cm)

COUNTRY/REGION OF ORIGIN Japan

COMMON COLORS

■ bay
■ chestnut
■ brown
□ gray
▨ roan

A native of Japan, the Dosanko is sometimes also called the Hokkaido, after the island in the north of the country from which it originates. Standing at around 13.2 hands (53 inches/134 cm), the Dosanko is a sturdy pony type, although its head is clearly of horse origin. It has light feathering and in all respects is a mini heavy or draft horse. The animal is used for both ridden and pack horse work in areas where motorized access is still difficult, reprising its earlier role.

The Dosanko is alone in being the only native equine breed in Japan that is not endangered. In 1973, a brood registry was founded, and numbers have varied from upward of 1,000 registered horses to a peak of around 3,000 in 1990. The Dosanko is hardy and tractable, and it is popular as a riding animal and trekking pony along tourist routes on the island of Hokkaido, where some of the breed do still live in the wild. It is said that the wild Dosanko will submerge itself into snow drifts, using them for insulation from the bitter winds and harsh temperatures that are a feature of the island during the winter.

Australian Waler

HEIGHT RANGE
14 to 16.2 hands
(56 to 65 in./142–165 cm)

COUNTRY/REGION OF ORIGIN Australia

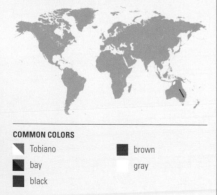

COMMON COLORS

◤ Tobiano
◼ bay
◼ black

◼ brown
◻ gray

This is an Australian breed of riding horse that originates from New South Wales, hence the name "Waler" or "Whaler"—both spellings are seen. Created from a variety of different breeds and types that were imported into the country in the nineteenth century and mixed with native stock, the distinct breed began to emerge as it was utilized for cavalry work, exploration expeditions, and as a stock horse. The requirements for speed and endurance combined with the deprivations of harsh conditions created a tough and courageous horse that varied in type depending on the purpose for which it was intended.

After World War II, this plain little bush horse began to fall out of favor as it became possible to access continental Warmbloods and Thoroughbreds. Then in 1971, the Australian Stock Horse Society was formed to save this indigenous breed. Many of the bloodlines were fairly modern and had outside influence, so in the mid-1980s, in an attempt to preserve the old Waler lines, the Waler Horse Society of Australia was formed, accessing stock in remote pockets of the outback that had been untouched by outside breeding influences. These breeders pride themselves in remaining completely true to the type of the old Australian Waler.

Don

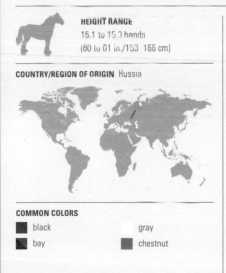

HEIGHT RANGE
15.1 to 15.3 hands
(60 to 61 in./153–155 cm)

COUNTRY/REGION OF ORIGIN Russia

COMMON COLORS

- black
- bay
- gray
- chestnut

So called after the Don River that runs through the Russian steppes, from where this horse originates, the Don is a riding horse that evolved from imported blood, such as Arab and Turkmenian, crossed with the feral horses of the steppes. Standing at 15.1 to 15.3 hands (60 to 61 inches/153–154 cm), the Don is rarely taller than 16 hands (64 inches/163 cm). It is often chestnut in color and is a quality muscular horse, although with a tendency to an upright shoulder, so it lacks the stride of horses with more Thoroughbred influence. However, its size and agility has made it popular as a cavalry remount, and it has immense qualities of endurance honed through decades of surviving in a combination of warm and windy summers and severe winters.

Numbers were decimated after the Russian Revolution and World War I but quickly rallied in the early twentieth century by the formation of several military studs and with the support of the Cossacks. Today, the Don is popular as a saddle horse and is often used in famous Cossack displays, where four horses are ridden together side by side with a sole horseman standing upright on the back of one of the horses. It is a fine example of a general riding horse with enough speed and athletic prowess for modern sport but also exceptional hardiness and endurance.

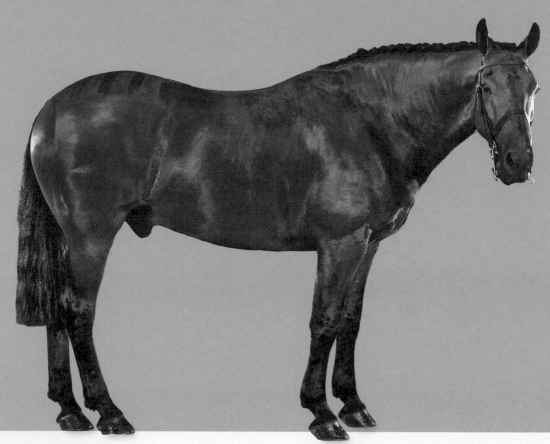

Cleveland Bay

HEIGHT RANGE
16 to 16.2 hands
(64 to 65 in./163–165 cm)

COUNTRY/REGION OF ORIGIN England

COMMON COLORS

■ bay

This old British breed of horse is characterized by its bay coloring and its association with the Cleveland area in the north of England. Originally a heavier breed, although never a true draft horse, the Cleveland Bay was used on the land until the arrival of the carriage in around the 1700s, when Thoroughbreds were used selectively to lighten the breed and they became a coach horse. The Cleveland Bay Horse Society was formed in 1884 to support dwindling numbers. The breed initially prospered, but it experienced huge losses during World War I. Numbers continued to fall and after World War II, there were few of the breed left. Fortunately, Her Majesty the Queen stepped in and bought a purebred colt called Mulgrave Supreme, born in 1961 and destined for export, which was made available to the public at stud and numbers began to rise.

The Cleveland Bay was a popular cross with the Thoroughbred during the 1960s and 1970s to produce a sport horse with speed, stamina, and bone. However, the rise of the modern Warmblood across postwar Europe saw numbers again begin to dwindle, because intensive Warmblood breeding programs produced a modern and seemingly more fashionable horse for equine sport. Today, the breed is listed by the Rare Breeds Survival Trust as critical with low numbers.

Criollo

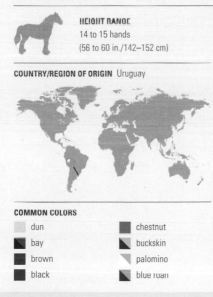

HEIGHT RANGE
14 to 15 hands
(56 to 60 in./142–152 cm)

COUNTRY/REGION OF ORIGIN Uruguay

COMMON COLORS

▢ dun		▨ chestnut	
◪ bay		◩ buckskin	
▨ brown		▧ palomino	
■ black		◪ blue roan	

The derivation of the word "Criollo" in both Spanish and Portuguese means "born in the Americas," and the Criollo horse is a native of Brazil, Paraguay, Argentina, and Uruguay. Originating from the Andalusian horse in the sixteenth century, this ancestry is still evident today in the conformation and appearance of the horse. Standing around 14 to 15 hands (56 to 60 inches/142–152 cm), the Criollo is a small, agile, and compact stock horse and the favored mount of the gaucho in the South American pampas. In the late nineteenth century, efforts were made to produce a larger animal by crossing the Criollo with the Thoroughbred, and the original animal was nearly lost until a breed registry was established in the twentieth century. There was much dispute and debate over what the breed standard should be, and it took several decades before this was established; during the 1930s, many Criollo horses who did not conform to this ideal were culled.

The Criollo has a reputation of being one of the toughest and most hardy horses in the world. In the last century, endurance feats were performed to test the ability of the breed to travel long distances bearing weight and with the minimum of maintenance. The Criollo is finding an increasing role today in the modern sport of endurance riding.

Holsteiner

HEIGHT RANGE
16 to 17 hands
(64 to 68 in./163–173 cm)

COUNTRY/REGION OF ORIGIN Germany

COMMON COLORS

- black
- bay
- brown
- chestnut
- gray

The Holsteiner, one of the Warmblood breeds that Germany is famous for, comes from the Schleswig-Holstein region in the north of the country. Many of the German Warmblood breeds are a real blend of bloodlines and breeds, but the Holsteiner represents a smaller, more select group of horses due to the closed policy of the Holsteiner Studbook. Despite this restriction on bloodlines, Holsteiner breeders have been determined to move with the times and created horses for different roles, always seeking to use whichever breed and bloodlines were necessary to eliminate unwanted traits, and in modern times they have produced a top class sport horse.

It is probably fair to say that the Holsteiner's success and popularity as a breed has been most prolific in show jumping rather than dressage. The Holsteiner has always been a jumping powerhouse known for a big engine and a high set on neck, and the breed has produced some of the most successful and well-known show jumpers the sport has seen. The influential sire, Ladykiller XX, needs no introduction and is a classic example of the use of a Thoroughbred to add refinement to a heavier frame with spectacular success. Ladykiller XX went on to produce Lord and Landgraf I, both hugely successful and influential horses.

Missouri Fox Trotter

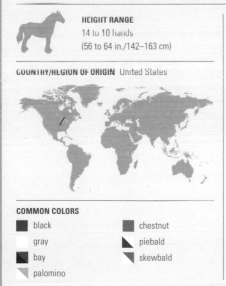

HEIGHT RANGE
14 to 16 hands
(56 to 64 in./142–163 cm)

COUNTRY/REGION OF ORIGIN United States

COMMON COLORS

- black
- gray
- bay
- palomino
- chestnut
- piebald
- skewbald

This is a breed of horse identified by its location, the U.S. state of Missouri, and for its ambling gait known as the fox trot. The usual trotting gait is a two-beat movement with the horse's legs moving in diagonal pairs and a moment of suspension in between. In contrast, the fox trot is a four-beat gait in which the front foot of the diagonal pair lands before the hind, eliminating the moment of suspension for a smoother ride. The horse must retain one foot on the ground at all times. This unusual gait was developed by early settlers in the nineteenth century, because the Missouri Fox Trotter was principally a stock horse, and this made it comfortable to spend long hours in the saddle over the rocky terrain. It also helped the horse's endurance performance with such an economical way of moving. The first breeders used other gaited horses, such as the Tennessee Walking Horse, but not to the exclusion of other influences such as Arab, Morgan, and the Standardbred.

A breed registry was opened in 1948 and now records the data of 100,000 horses; a European Registry began in 1992. The Trotter is still used for ranch work and trail riding, and they are popular with disabled riders due to the smoothness of the gait.

Morgan

HEIGHT RANGE
14.2 to 15.2 hands
(57 to 61 in./144–154 cm)

COUNTRY/REGION OF ORIGIN United States

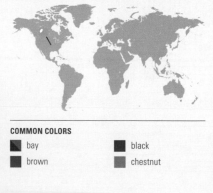

COMMON COLORS

- ■ bay
- ■ brown
- ■ black
- ■ chestnut

The Morgan horse is an American breed. Every single horse can be traced back to one original colt named Figure, born around 1790 in Massachusetts and owned by a man called Justin Morgan, hence the name of the breed. There has been much conjecture about the breeding of this first colt and certainly, if you look at the Morgan, it is possible to see different influences, such as Arab, Thoroughbred, Barb, Friesian, and even Welsh. The Morgan became popular in the United States in the 1850s as a quality horse with clean limbs, an expressive and beautiful head, and plenty of presence. It is a relatively compact horse with a fairly high action, which is perhaps a nod toward Welsh influences. The Morgan has formed part of the foundation for other American breeds, such as the Standardbred and the Tennessee Walking Horse. Before mechanization, the Morgan was the choice of the American Army as a soldiers' mount and is still used as a police horse in some American states.

With the average height of a Morgan at about 15 hands (60 inches/152 cm), this hardy and versatile horse is popular across many equestrian disciplines. It has enough quality for riding horse and hack classes but also the quickness and stamina for other equestrian disciplines. Its size makes it a popular mount for children and adults alike.

Mustang

HEIGHT RANGE
13 to 15 hands
(52 to 60 in./132–152 cm)

COUNTRY/REGION OF ORIGIN United States

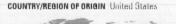

COMMON COLORS

■ bay	■ gray
■ black	◪ Appaloosa
■ chestnut	◩ Buckskin
◺ palomino	■ Cremello

The Mustang is the free-roaming feral horse that is synonymous with the pioneering spirit of the Wild West. The word "Mustang" has Spanish roots and the Mustang horse has certainly been influenced by earlier horses brought to the United States by the Spanish colonialists—so these feral horses are descended from domesticated animals. The fortunes of the Mustang have ebbed and flowed largely because of humans, who at times have let it be and at other times, helped himself to the population when needed, for example, during the Spanish–American War in the late 1800s.

Ethnic origin varies and depends to some degree on the location of the specific herd. There are 72,000 Mustangs currently in the country, and more than half of them are found in Nevada, with other large clusters in California, Utah, Montana, Oregon, and Wyoming. Free-roaming Mustangs are now managed by the Bureau of Land Management, which also attempts to control the population by rounding up herds and offering horses to private individuals. These are contained in holding areas and there is estimated to be some 45,000 horses currently in holding facilities. Some ultimately do end up in the food chain, because there are not enough people coming forward to claim them.

Tennessee Walker

HEIGHT RANGE
14.3 to 17 hands
(57 to 68 in./145–173 cm)

COUNTRY/REGION OF ORIGIN United States

COMMON COLORS

- bay
- black
- chestnut
- dun
- pinto

An American gaited horse, the Tennessee Walker has a unique four-beat walk that, in essence, runs—it is a breed that is never far from controversy. The Tennessee Walking Horse Breeders' Association was founded in 1935 to recognize a breed developed in the previous century, when gaited American horses were crossed with Spanish Mustangs. The name was changed in 1974 to the Tennessee Walking Horse Breeders and Exhibitors Association to reflect the interest in showing the horses. The studbook was closed in 1947, so every foal since must have both parents registered to be eligible for inclusion.

The horse has in recent times attracted attention from welfare organizations who dislike the practice of shoeing some Tennessee Walker's with "stacks," which are built-up pads designed to exaggerate or enhance the horse's existing gait. The 1970 Horse Protection Act has sought to criminalize some other practices, such as "soring" (making the legs sore), which is designed to exaggerate its gait for performance purposes but is considered abuse of the horse. This has led to division within the Walker community and the formation of different breed organizations. Not all horses are shown like this; some remain "flat-shod," with a conventional horse shoe and a less exaggerated gait.

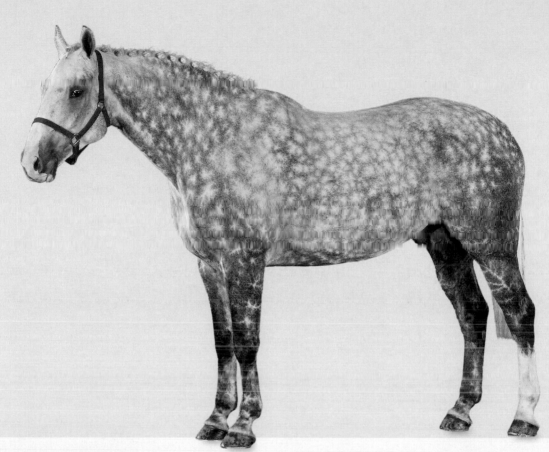

Irish Sport Horse

HEIGHT RANGE
15 to 17.2 hands
(60 to 69 in./152–175 cm)

COUNTRY/REGION OF ORIGIN Ireland

COMMON COLORS

■ chestnut	■ gray
■ bay	■ dun
■ brown	■ dapple gray

The Irish Sport Horse is a relatively modern concept, usually the combination of a Thoroughbred and an Irish Draft. Some years ago, a horse of mixed breeding was called Irish Draft x Thoroughbred or three-quarters bred or seven-eighths bred (the Thoroughbred element), with the remaining part being the Irish Draft blood. There has been a modern tendency to describe these horses as Irish Sport Horses, and this name covers horses of established and known breeding, a mix of Thoroughbred, Irish Draft, and Irish Cob or Pony, and, more lately, Warmblood breeding or horses whose breeding is partly or completely unknown. So the term Irish Sport Horse can refer to a horse that has come out of Ireland, but its breeding is unknown albeit in appearance it can be clearly seen to be of Thoroughbred or Irish Draft descent.

The Irish Sport Horse Studbook is now an umbrella for many different breeding influences including Warmbloods. The studbook includes a separate category for the "Traditional Irish Horse"; however, a horse of unknown breeding will be described on its passport as an Irish Sport Horse. It is a breed title that requires further investigation.

Konik

HEIGHT RANGE
12.3 to 13.3 hands
(49 to 53 in./125–135 cm)

COUNTRY/REGION OF ORIGIN Poland

COMMON COLORS
dun

The Konik is a primitive horse—the name means, "small horse" in Polish. The primitive status of this breed, derived from the now-extinct Neolithic tarpan, is visible by the zebra striping on the front and hind legs and the dorsal or eel stripe, which runs along the center of the back from wither to tail. The ponies have a mouse dun coat color and a two-tone mane.

During World War II, German zoo directors in Berlin and Munich were supported by senior members of the government in their efforts to re-create the primitive tarpan, the last one of which had died in Europe in 1910, rendering the breed extinct. The tarpan horse is a major feature in German folklore, part of the story of the German forests and hunting culture. Entire herds of Koniks were transported from Poland into Germany, and those remaining after the war were then returned to national parks in Soviet-occupied Poland. When the Iron Curtain finally fell, western European countries were able to make use of this breed for conservation grazing projects across Europe, including the United Kingdom.

Comtois

HEIGHT RANGE
14 to 16 hands
(56 to 64 in./142–163 cm)

COUNTRY/REGION OF ORIGIN France

COMMON COLORS
■ chestnut

The Comtois is a French breed originating from the area in eastern France that borders Switzerland, now called the Franche-Comté. A small draft horse, it starts at around 14 hands (56 inches/142 cm) and grows no bigger than about 16 hands (64 inches/ 163 cm). It is mostly chestnut, although it can vary from a bright golden color to a dark bay or chocolate brown. The eye-catching mane and tail are a lighter flaxen color that creates a striking effect. Prior to World War I until after World War II, the breed lent itself to agricultural work and was a popular farm horse. Historically, the Comtois was the classic mount of the soldier and favored by Napoleon in his Russian campaign.

Today, the breed still hauls timber in the high pine forests found in its native region, but it has also developed popularity as a riding horse. Being quicker and more agile than other heavy horses and also trainable, good natured, and hard working, the breed is finding increasing popularity under saddle. The Comtois is a horse that can easily manage agricultural work around the farm or other duties in harness but yet also works under saddle with ease. It is not unusual to see one now in harness in trade turnout classes at the big shows or under saddle in the breed sections.

Friesian

HEIGHT RANGE
14.2 to 17 hands
(57 to 68 in./144–173 cm)

COUNTRY/REGION OF ORIGIN Netherlands

COMMON COLORS
■ black

If you have seen a horse-drawn funeral cortege with black horses, then the hearse was probably pulled by Friesian horses. Originally a light draft horse, the size and color of this eye-catching horse make it a perfect choice for a funeral procession. Originating from Friesland in the Netherlands, the breed has followed the usual historical progression for one of this type from war-horse in medieval times to working, agricultural horse before standing aside for the progress of the Industrial Revolution and the mechanization of farming. It is a striking and flamboyant horse with an agile, nimble movement due in part to an infusion of Spanish blood in the sixteenth century.

Offering a movement that is perhaps freer and cleaner than some other comparable breeds and types, the Friesian has begun to find favor among dressage riders. Interestingly, it has also become a favorite choice as a movie horse for period dramas and battle re-creations, because, again, the conformation and vibrancy of the breed appeal to the often romanticized interpretation of the moviemaker. Usually black in color with only the smallest amount of white hair permitted, the Friesian commonly stands around 15.3 hands (61 inches/155 cm), but it is a proud and upstanding horse and can often appear bigger.

Haflinger

HEIGHT RANGE
13.3 to 14.3 hands
(53 to 57 in./135–145 cm)

COUNTRY/REGION OF ORIGIN Austria

COMMON COLORS
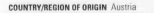 chestnut (all shades)

The Haflinger has something of a niche yet cult following throughout the world. This diminutive little horse, most definitely not a pony, can trace its lineage back to one stallion, Folie 249, who was born in 1874. Folie 249 was a cross between an Arab stallion and a Tyrolean mountain mare thought to be of Warmblood origin. The Haflinger was originally used as a light pack horse in the mountains, a role for which it was ideally suited. The North Tyrolean Breeding Association was formed in 1921, and in 1946 a breeding program was begun to take the Haflinger back to its original conformation, a horse of a lighter frame, which had been sidelined during the two world wars, when there were more pressing matters to attend to other than selective horse breeding.

The Haflinger combines the quickness and surefooted attributes of a pony with the strength and frame of a small horse. It has become a popular mount under saddle, and not least because of its striking color. The breed embraces all shades of chestnut, from a pale golden color to a dark liver chestnut. The mane and tail should be light or white in contrast, and the breed standard prefers a minimum of white markings confined to the head. A versatile horse, its equable temperament makes it a popular choice for riding, and its hardy constitution enables it to thrive on limited rations.

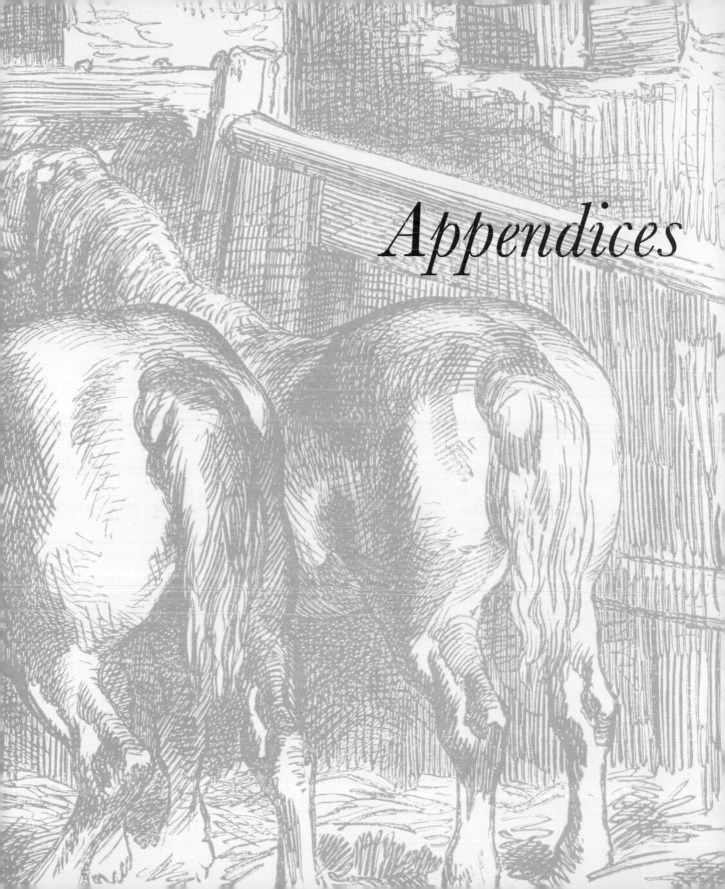

Appendices

Bibliography ⌀

CHAPTER 1

HOLDERNESS-RODDAM, J. (1999). *The Life of Horses.* Octopus Publishing Group Ltd. London.

GOTO, H., RYDER, O.A., FISHER, A.R., SCHULTZ, B., POND, S.L., KOSAKOVSKY., NEKRUTENKO, A., MAKOVA, K.D. (2011). *A Massively Parallel Sequencing Approach Uncovers Ancient Origins and High Genetic Variability of Endangered Przewalski's Horses.* Genome Biology and Evolution. 3: 1096–1106. doi:10.1093/gbe/evr067.

SHAH, N. (2002). *Status and action plan for the kiang (Equus kiang).* In: P. D. Moehlman (ed.), Equids: Zebras, Asses, and Horses. Status Survey and Conservation Action Plan, pp. 72–81. Switzerland.

KING, S.R.B. & MOEHLMAN, P.D. (2016). *Equus quagga.* IUCN Red List of Threatened Species. Version 2015.1. International Union for Conservation of Nature.

ORLANDO, L., GINOLHAC, A. L., ZHANG, G., FROESE, D., ALBRECHTSEN, A., STILLER, M., SCHUBERT, M., CAPPELLINI, E., PETERSEN, B., MOLTKE, I., JOHNSON, P. L. F., FUMAGALLI, M., VILSTRUP, J. T., RAGHAVAN, M., KORNELIUSSEN, T., MALASPINAS, A. S., VOGT, J., SZKLARCZYK, D., KELSTRUP, C. D., VINTHER, J., DOLOCAN, A., STENDERUP, J., VELAZQUEZ, A. M. V., CAHILL, J., RASMUSSEN, M., WANG, X., MIN, J., ZAZULA, G. D., SEGUIN-ORLANDO, A., MORTENSEN, C. (2013). *Recalibrating Equus evolution using the genome sequence of an early Middle Pleistocene horse.* Nature. 499 (7456): 74–78. doi:10.1038/nature12323.

WENDLE, JOHN. (2016). *Animals Rule Chernobyl 30 Years After Nuclear Disaster.* National Geographic. Retrieved May 2, 2016.

MACFADDEN, B. J. (1984). *Astrohippus and Dinohippus.* J. Vert. Paleon. 4 (2): 273–283. doi:10.1080/02724634.1984.10012009.

HEDGE, J., and WAGONER, D.M. (2004). *Horse Conformation: Structure, Soundness and Performance.* Guilford, CT: Globe Pequot. pp. 307–308. ISBN 1-59228-487-6. OCLC 56012597.

MACFADDEN, B. J. (1984). *Systematics and phylogeny of Hipparion, Neohipparion, Nannippus, and Cormohipparion (Mammalia, Equidae) from the Miocene and Pliocene of the New World.* Bulletin of the American Museum of Natural History. 179 (1): 1–195.

SALESA, M. J.; SÁNCHEZ, I. M. & MORALES, J. (2004). *Presence of the Asian horse Sinohippus in the Miocene of Europe.* Acta Palaeontologica Polonica. 49 (2): 189–196.

PROTHERO, D. R. and SHUBIN, N. (1989). *The evolution of Oligocene horses.* The Evolution of Perissodactyls. pp. 142–175. New York: Clarendon Press.

GRUBB, P. (2005). *Order Perissodactyla.* In WILSON, D.E.; REEDER, D.M. *Mammal Species of the World: A Taxonomic and Geographic Reference* (3rd ed.). Johns Hopkins University Press. p. 630–631. ISBN 978-0-8018-8221-0.

CHAPTER 2

BUDRAS, K.D., SACK, W.O., ROCK, S., HOROWITZ, A. and BERG, R. (2009) *Anatomy of the Horse.* 5th Ed. Schlutersche Verlagsgesellschaft mbH & Co. Hannover, Germany.

TAYLOR, S. (2015) *A review of equine sepsis.* Equine Veterinary Education. 27,2. 99–109. https://dairy.ahdb.org.uk/market-information/farming-data/milk-yield/average-milk-yield/#.WX_LWYTyvIU

RIEGAL, R.J. and HAKOLA, S.E. (1999) *Illustrated atlas of clinical equine anatomy and common disorders of the horse.* EQUUS Magazine. USA.

"Horse Nutrition - Diet Factors - Water." Bulletin 762-00, Ohio State University. Web site accessed August 2017. http://arquivo.pt/wayback/20090708015300/http://ohioline.osu.edu/b762/b762_6.html

HEFFNER, H.E. and HEFFNER, R.S. (2005) *Sound localization and high-frequency hearing in horses.* The Journal of the Acoustical Society of America. 73, S1. DOI: 10.1121/1.2020377

LEE, J.R., HONG, C.P., MOON, J.W., JUNG, Y.D., KIM, D.S., KIM, T.H., GIM, J.A., BAE, J.H., CHOI, Y., EO, J., KWON, Y.J., SONG, S., KO, J., YANG, Y.M., LEE, H,K., PARK, K.D., AHN, K., DO, K.T., HA, H.S., HAN, K., YI, M.J., CHA, H.J., CHO, B.W., BHAK J., and KIM, H.S. (2014) *Genome-wide analysis of DNA methylation patterns in horse.* BMC Genomics. 15 (1): 598.

ALLISON, L., PENG, L., GOTO, H., CHEMNICK, L., RYDER, O.A., MAKOVA, K.D. (2009) *Horse Domestication and Conservation*

Genetics of Przewalski's Horse Inferred from Sex Chromosomal and Autosomal Sequences. Mol. Biol. Evol. 26 (1): 199–208. PMID 18931383. doi:10.1093/molbev/msn239.

KEANE, M., PAUL, E., STURROCK, C., RAUCH, C. and RUTLAND, C.S. (2017) *Computed tomogaphy in veterinary medicine: currently published and tomorrow's vision*. Computed Tomography - Advanced Applications. InTechOpen. https://www.intechopen.com/books/computed-tomography-advanced-applications/computed-tomography-in-veterinary-medicine-currently-published-and-tomorrow-s-vision.

SIMSON, E., RUTLAND, R and RUTLAND, CS. (2017) *Genomic insights into cardiomyopathies: a comparative cross-species review.* Veterinary Sciences. 4 (1), 19. doi:10.3390/vetsci4010019.

METALLINOS, D.L., BOWLING, A.T., and RINE, J. (1998) *A missense mutation in the endothelin-B receptor gene is associated with Lethal White Foal Syndrome: an equine version of Hirschsprung Disease.* Mammalian Genome. New York: Springer New York. 9 (6): 426–31. PMID 9585428. doi:10.1007/s003359900790.

CASTLE, W.E. (1948) *The Abc of Color Inheritance in Horses.* Genetics. 33 (1): 22–35. PMC 1209395 . PMID 17247268.

LODATO, S., and ARLOTTA, P. (2015) *Generating Neuronal Diversity in the Mammalian Cerebral Cortex.* Annual Review of Cell and Developmental Biology. 31 (1): 699–720. doi:10.1146/annurev-cellbio-100814-125353.

LUI, J. H., HANSEN, D. V., and KRIEGSTEIN, A. R. (2011). *Development and Evolution of the Human Neocortex.* Cell. 146 (1): 18–36. doi:10.1016/j.cell.2011.06.0

KOVAC M, LITVIN YA, ALIEV RO, ZAKIROVA EY, RUTLAND CS, KIYASOV AP and RIZVANOV AA (2017) *Gene Therapy Using Plasmid DNA Encoding Vascular Endothelial Growth Factor 164 and Fibroblast Growth Factor 2 Genes for the Treatment of Horse Tendinitis and Desmitis: Case Reports.* Front. Vet. Sci. 4:168. doi: 10.3389/fvets.2017.00168

BREGA, J. (2005) *Anatomy and Physiology.* J.A. Allen. London, UK.

KOVAC M, LITVIN YA, ALIEV RO, ZAKIROVA EY, RUTLAND CS, KIYASOV AP and RIZVANOV AA (2018) *Gene Therapy Using Plasmid DNA Encoding VEGF164 and FGF2 Genes: A Novel Treatment of Naturally Occurring Tendinitis and Desmitis in Horses.* Front. Pharmacol. 9:978. doi: 10.3389/fphar.2018.00978

CHAPTER 3

Books

BALDWIN. J. D. and *Baldwin, J. I.* (2001) *Behavior Principles in everyday life.* New Jersey. Prentice Hall.

BROOM, D. M., & FRASER, A. F. (2015), *Domestic animal behaviour and welfare.* Cabi.

BUDIANSKY, S. (1997) *The nature of horses.* Cambridge. Free Press

CARLSON, N. R. (1994), *Physiology of behavior.* Allyn & Bacon.

FRASER, A. F. (1992). *The behaviour of the horse.* CAB international.

McDONNELL, S, (2003) *The equid ethogram: A practical field guide to horse behaviour.* Eclipse press. London.

McGREEVY, P. (2013) *Equine behaviour, a guide for veterinarians and equine scientists.* 2nd edn. London Elsevier

MILLS, D. and McDONNELL S. (2005) *The domestic horse; the evolution, development and management of its behaviour.* Cambridge. University Press.

MILLS, D. S., & NANKERVIS, K. J. (2013). *Equine behaviour: principles and practice.* John Wiley & Sons.

PANKSEPP J. (1998) *Affective Neuroscience: The Foundations of Human and Animal Emotions.* Oxford University Press, New York.

RANSOM, J. I., & KACZENSKY, P. (Eds.). (2016). *Wild Equids: Ecology, Management, and Conservation.* JHU Press.

ROSSDALE, P. D. (1975). *The horse-from conception to maturity.*

Journals

ANDRIEU, J., HENRY, S., HAUSBERGER, M., & THIERRY, B. (2016). Informed horses are influential in group movements, but they may avoid leading. Animal cognition, 19(3), 451-458.

BEKOFF, M. (2000). Animal Emotions: Exploring Passionate Natures: Current interdisciplinary research provides compelling evidence that many animals experience such emotions as joy, fear, love, despair, and grief - we are not alone. BioScience, 50(10), 861-870.

BERTONE, J. J., FURR, M., & REED, S. (2015) Sleep and Sleep Disorders in Horses. Equine Neurology, Second Edition, 123-129.

BLISS, T. V., & COLLINGRIDGE, G. L. (1993). A synaptic model of memory: long term potentiation in the hippocampus. Nature, 361(6407), 31.

BOISSY, A., & LEE, C. (2014). How assessing relationships between emotions and cognition can improve farm animal welfare. Revue scientifique et technique (International Office of Epizootics), 33, 103-110.

BRIEFER, E. F., MAIGROT, A. L., MANDEL, R., FREYMOND, S. B., BACHMANN, I., & HILLMANN, E. (2015). Segregation of information about emotional arousal and valence in horse whinnies. Scientific reports, 4, 9989.

CREGIER, S.E. (2009) Best practices: surface transport of the horse. Available at: https://www.academia.edu/6616417/Best_Practices_in_Horse_Transport_Animal_Transportation_Association_Proceedings

EKMAN, P. (1992). An argument for basic emotions. Cognition & emotion, 6(3-4), 169-200.

GAUNITZ, C., FAGES, A., HANGHØJ, K., ALBRECHTSEN, A., KHAN, N., SCHUBERT, M., & DE BARROS DAMGAARD, P (2018) Ancient genomes revisit the ancestry of domestic and Przewalski's horses. Science, 360 (6384), 111-114.

HALL, C., GOODWIN, D., HELESKI, C., RANDLE, H., & WARAN, N. (2008). Is there evidence of learned helplessness in horses?. Journal of Applied Animal Welfare Science, 11(3), 249-266.

HANGGI, E. B. (2005, December). The thinking horse: cognition and perception reviewed. In AAEP Proceedings (Vol. 51, pp. 246 255).

KYDD, E., PADALINO, B., HENSHALL, C., & McGREEVY, P. (2017). An analysis of equine round pen training videos posted online: Differences between amateur and professional trainers. PloS one, 12(9), e0184851.

LEDOUX, J. E., MOSCARELLO, J., SEARS, R., & CAMPESE, V. (2017). The birth, death and resurrection of avoidance: a reconceptualization of a troubled paradigm. Molecular psychiatry, 22(1), 24.

McGREEVY, P. D. (2007) The advent of equitation science. Veterinary Journal, 174, pp. 492–500

McGREEVY, P. D., & McLEAN, A. N. (2007). Roles of learning theory and ethology in equitation. Journal of Veterinary Behavior: Clinical Applications and Research, 2(4), 108-118.

McLEAN, A. N., & CHRISTENSEN, J. W. (2017). The application of learning theory in horse training. Applied Animal Behaviour Science, 190, 18-27.

RUSSELL, J. A. (2003). Core affect and the psychological construction of emotion. Psychological review, 110(1), 145.

WATHAN, J., PROOPS, L., GROUNDS, K., & MCCOMB, K. (2016). Horses discriminate between facial expressions of conspecifics. Scientific reports, 6, 38322.

WARMUTH, V., ERIKSSON, A., BOWER, M. A., BARKER, G., BARRETT, E., HANKS, B. K., & SOYONOV, V. (2012). Reconstructing the origin and spread of horse domestication in the Eurasian steppe. Proceedings of the National Academy of Sciences, 109(21), 8202-8206.

Websites

Equine Behaviour and Training Association http://www.ebta.co.uk/

CHAPTER 4

Books

ADELMAN, M., & THOMPSON, K. (Eds.). (2017). *Equestrian cultures in global and local contexts.* Springer.

McCAFFREY, A. (2002). *The Lady.* Ballantine Books

EDWARDS, E. H., LANGRISH, B., & HOUGHTON, K. (1994). *The encyclopedia of the horse.* London: Dorling Kindersley.

JEFFCOTT, L. B., KIDD, J. A., & BAINBRIDGE, D. (2017). The Normal Anatomy of the Osseous and Soft Tissue Structures of the Back and Pelvis. Equine Neck and Back Pathology: Diagnosis and Treatment, 9.

NYLAND, A. (1993). *The Kikkuli method of horse training.* Kikkuli Research Publications.

RAULFF, U. (2017). *Farewell to the Horse: the final century of our relationship.* Penguin UK.

WILLIAMS, W. (2015). *The Horse: The Epic History of Our Noble Companion.* Macmillan.

XENOPHON, (430–350 BC) *The art of horsemanship.* MORGAN M. H. (ed, trans 1893), New York: Dover (2006)

Journals

BROWN, S. M., & CONNOR, M. (2017). Understanding and Application of Learning Theory in UK-based Equestrians. Anthrozoös, 30(4), 565-579.

BIRKE, L., and HOCKENHULL, J. (2015) 'Journeys Together: Horses and Humans in Partnership'. Society and Animals, 23(1), pp. 81-100.

CONNORS, S. & FELDMAN, L. (2009) The Equine Industry as a Global Market. Available at: https://www.researchgate.net/publication/276830364_The_Equine_Industry_as_a_Global_Market

DASHPER, K. (2017). Listening to horses. Society & Animals, 25(3), 207-224.

HAUSBERGER, M., ROCHE, H., HENRY, S., and VISSER, E. K. (2008). 'A review of the human–horse relationship'. Applied Animal Behaviour Science, 109(1), pp. 1-24.

KOENEN, E. P. C., ALDRIDGE, L. I., and PHILIPSSON, J. (2004). An overview of breeding objectives for warmblood sport horses. Livestock Production Science, 88(1), pp. 77-84

McGREEVY, P. D. (2007) The advent of equitation science. Veterinary Journal, 174, pp. 492–500

McGREEVY, P. D., & McLEAN, A. N. (2007). Roles of learning theory and ethology in equitation. Journal of Veterinary Behavior: Clinical Applications and Research, 2(4), 108-118.

McGREEVY, P. et al., (2018). Using the Five Domains Model to Assess the Adverse Impacts of Husbandry, Veterinary, and Equitation Interventions on Horse Welfare. Animals, 8(3), 41.

McLEAN, A. N., & CHRISTENSEN, J. W. (2017). The application of learning theory in horse training. Applied Animal Behaviour Science, 190, 18-27.

MELLOR, D. J., & BEAUSOLEIL, N. J. (2015). Extending the 'Five Domains' model for animal welfare assessment to incorporate positive welfare states. Animal Welfare, 24(3), 241-253.

WALKER, J. E. (2005). 'To amaze the people with pleasure and delight': an analysis of the horsemanship manuals of William Cavendish, first Duke of Newcastle (1593-1676) (Doctoral dissertation, University of Birmingham).

WEBSTER, J. (2016). Animal welfare: Freedoms, dominions and "a life worth living". Animals, 6(6), 35.

Websites

Brooke Action for Working Horses and Donkeys https://www.thebrooke.org/

Equine Business Association https://www.equinebusinessassociation.com/equine-industry-statistics/

European Horse Network http://www.europeanhorsenetwork.eu/

International Society for Equitation Science https://equitationscience.com/

Federation Equestre Internationale https://www.fei.org/

World Horse Welfare http://www.worldhorsewelfare.org/Home

World Organisation for Animal Health (OIE) http://www.oie.int/

CHAPTER 5

Books

HENDRICKS, B. L. (2007). *International encyclopedia of horse breeds.* University of Oklahoma Press.

SWINNEY, N. J. (2006). *Horse Breeds of the World.* Hamlyn.

Index ∽

Author Biographies

Dr Catrin Rutland is an Associate Professor of Anatomy and Developmental Genetics at The University of Nottingham. She has a BSc (Hons), MSc and PhD and also teaching qualifications PGCHE and MMedSci (Medical education). Her research involves cardiovascular work, which includes understanding both the heart and blood vessels in a number of species ranging from humans to horses. Her equine and donkey research concentrates on the anatomy and physiology of the limbs and lameness, and also on gene therapy as a treatment. In addition to her research publications and teaching at the vet school, Catrin also writes articles/chapters for magazines, newspapers and books, and writes papers and gives talks aimed at young people and the general public.

Debbie Busby MSc MBPsS is a behaviourist and international speaker and author. Debbie has a BSc in Psychology and gained her MSc in Applied Animal Behaviour and Welfare at Newcastle University. She is a full member of the Association of Pet Behaviour Counsellors and is registered with the Animal Behaviour and Training Council as a Clinical Animal Behaviourist.

Debbie's consultancy Evolution Equine Behaviour is based in the UK and she consults internationally on serious behavioural disorders in horses. Debbie delivers presentations and workshops on all aspects of equine behaviour, consulting and human behaviour change, and she is a co-author of Equine Behaviour in Mind: Applying Behavioural Science to the Way We Keep, Work and Care for Horses (2018) published by 5M Publishing.

Acknowledgments ◌

Debbie Busby would like to thank Kelly Taylor-Saunders, Jenni Nellist, Anne Howard, Michelle Wilson and Linda Smith for their valuable research contributions.

The publisher would like to thank the following for permission to reproduce copyright material:

Alamy/Aflo Co. Ltd.: 145T; Agencja Fotograficzna Caro: 145B; Arco Images GmbH: 45TR; Arctic Images: 81; Bill Emrich: 143T; blickwinkel: 14; Blue Jean Images: 76; catnap: 210; Central Historic Books: 127R; Corbin17: 13; Juniors Bildarchiv GmbH: 202, 207; dpa picture alliance archive: 141B; Graham Prentice: 142; Greenshoots Communications: 109; Heritage Image Partnership Ltd : 117; Historic Images: 128; Interfoto : 28T; Jipen: 195; Juniors Bildarchiv GmbH: 85, 95T 141T, 155; Lanmas: 114; Lourens Smak: 44; Nature Picture Library: 140; Pavel Filatov: 28B; Prisma Archivo: 119; Reimar: 118B; Richard Tadman: 154; Rosanne Tackaberry: 110; The Print Collector : 64; Tomasz Wojnicz: 90; Vintage Archive: 149T.

Biodiversity Heritage Library/ Smithsonian Libraries: 17.

Bob Langrish: 102, 160T, 161B, 161T, 184, 205, 209.

DK Images/Bob Langrish: 164T, 177, 179, 180, 185, 194, 200, 201.

FLPA/Sabine Schwerdtfeger/ Tierfotoagentur: 193.

Getty Images/AFP/Alexander Klein: 160B; De Agostini/Bob Langrish: 189; Hulton Archive: 158; Mike Hewitt: 136; National Geographic Creative/Sisse

Brimberg: 115; UIG/Werner Forman: 132B.

H. Zell/Wikimedia/CC BY-SA 3.0: 16.

iStock/Nicoolay: 1.

Ivy Press/Andrew Perris: 166, 167, 168, 169, 170, 171, 172, 173, 174, 175, 181, 186, 187, 190, 192, 196, 198, 204, 208, 211, 213, 214, 215

Library of Congress/Detroit Publishing Company photograph collection: 144.

Mary Evans Picture Library: 8B, 9B; Tony Boxall: 9.

Shutterstock/4thebirds: 80; Abir Roy Barman: 112; Abramova Kseniya: 37BL, 139, 182, 203; alfernec: 74B; AlinArt: 122T; Anastasiia Golovkova: 29 (icons); Anastasija Popova: 104B; Andrey N Bannov: 57; andrey oleynik: 162T; anjajuli: 72; bcostelloe: 20; Beck Dunn Photography: 108B; Bildagentur Zoonar GmbH: 7; Buffy1982: 93; Callipso: 47, 68B; Chepko Danil Vitalevich: 84; Christian Mueller: 152B; cornfield: 79BL; cynoclub: 123T; E. Spek: 79BR; EQRoy: 25; Eric Isselee: 2, 3TR, 3CR, 3L, 37BR, 75C, 176, 183, 188; esherez: 150B; Everett Historical: 133T; fatir29: 88; Fedor Selivanov: 113; frankazoidtrvl: 96; GeptaYs: 36B, 148T; Gina Stef: 125BL; Govorov Evgeny: 70B; Grigorita Ko: 129B; Groomee: 106, 130; Hein Nouwens: 70T, 118T, 133B; hofhauser: 26; horsemen: 206; J. Marijs: 212; jacotakepics: 199; jakelv7500: 107; Jorge Maricato: 92; Juha Saastamoinen: 39TR; Julia Siomuha: 97; Katho Menden: 82B; kongsak sumano: 21T; Konstantin Tronin: 61; Kraft_Stoff: 148–9B; kryzhov: 147B; LanaG: 101;

Lenkadan: 38BL; Lenkadan: 197, 19B; Makarova Viktoria: 38BR; Manekina Serafima: 18T; mariait: 5B, 123B; marina eno 1: 100B; maziarz: 8T; Melinda Nagy: 45TL; Melory: 62; Mick Atkins: 147T; Mogens Trolle: 23B; Morphart Creation: 15, 18B, 66T, 120, 132T, 217; Mumemories: 98B; Nadzeya Shanchuk: 166–215 (icons); Nate Allred: 53T; Nigel Jarvis: 143B; Oleksandr Umanskyi: 83; Olga_i: 27T; paula french: 23T; Pavlina Trauskeova: 31; Pegasene: 129T; Perry Correll: 137B; Peter Etchells: 40; Petri Volanen: 39TL; pfluegler-photo: 33C, 33B; photo-denver: 138; pirita: 103; Robynrg: 74T; S. Bonaime: 178; salajean: 151; Sanit Fuangnakhon: 125BR; Sergei25: 19T; Sergey Kohl: 4, 104T; SF photo: 27B; Studio Barcelona: 41; SunnyMoon: 156; superjoseph: 150T; Svetlana Zhukova: 34; TasfotoNL: 126; tashh1601: 146; Tony Gatlin: 87; Vaclav Volrab: 21B; Vector Tradition: 5T, 78; vectorlight: 29 (map); Vladimir Melnik: 22; wavebreakmedia: 152T; Willyam Bradberry: 10; Wolf Avni: 35; Worraket: 75B; Zuzule: 95B.

Wellcome Collection/CC BY 4.0: 121, 124, 134, 135, 137T.

Wikimedia/CC-PD-Mark: 159, 127BR; Osama Shukir Muhammed Amin FRCP(Glasg)/CC BY-SA 4.0: 122B; Scherer/CC-PD-Mark: 24.

All reasonable efforts have been made to trace copyright holders and to obtain their permission for the use of copyright material. The publisher apologizes for any errors or omissions and will gratefully incorporate any corrections in future reprints if notified.